ON-SITE GUIDE
BS 7671:2008(2011)

IET Wiring Regulations 17th Edition
Incorporating Amendment No 1

IET Wiring Regulations Seventeenth Edition
BS 7671:2008(2011) Requirements for Electrical Installations

Published by The Institution of Engineering and Technology, London, United Kingdom

The Institution of Engineering and Technology is registered as a Charity in England & Wales (no. 211014) and Scotland (no. SC038698).

The Institution of Engineering and Technology is the new institution formed by the joining together of the IEE (The Institution of Electrical Engineers) and the IIE (The Institution of Incorporated Engineers). The new Institution is the inheritor of the IEE brand and all its products and services, such as this one, which we hope you will find useful.

The paper used to print this publication is made from certified sustainable forestry sources.

© 1992, 1995, 1998, 2002, 2004 The Institution of Electrical Engineers
© 2008, 2011 The Institution of Engineering and Technology

First published 1992 (0 85296 537 0)
Reprinted (with amendments) May 1993
Reprinted (with amendments to Appendix 9) July 1993
Reprinted (with amendments) 1994
Revised edition (incorporating Amendment No. 1 to BS 7671:1992) 1995
Reprinted (with new cover) 1996
Revised edition (incorporating Amendment No. 2 to BS 7671:1992) 1998
Second edition (incorporating Amendment No. 1 to BS 7671:2001) 2002 (0 85296 987 2)
Reprinted (with new cover) 2003
Third edition (incorporating Amendment No. 2 to BS 7671:2001) 2004 (0 86341 374 9)
Fourth edition (incorporating BS 7671:2008) 2008 (978-0-86341-854-9)
Reprinted (with amendments) October 2008
Fifth edition (incorporating Amendment No. 1 to BS 7671:2008) 2011 (978-1-84919-287-3)
Reprinted 2012
Reprinted (with minor corrections) 2013

This publication is copyright under the Berne Convention and the Universal Copyright Convention. All rights reserved. Apart from any fair dealing for the purposes of research or private study, or criticism or review, as permitted under the Copyright, Designs and Patents Act, 1988, this publication may be reproduced, stored or transmitted, in any form or by any means, only with the prior permission in writing of the publishers, or in the case of reprographic reproduction in accordance with the terms of licences issued by the Copyright Licensing Agency. Enquiries concerning reproduction outside those terms should be sent to the publishers at The Institution of Engineering and Technology, Michael Faraday House, Six Hills Way, Stevenage, SG1 2AY, United Kingdom.

Copies of this publication may be obtained from:
PO Box 96, Stevenage, SG1 2SD, UK
Tel: +44 (0)1438 767328
Email: sales@theiet.org
www.electrical.theiet.org/books

While the author, publisher and contributors believe that the information and guidance given in this work are correct, all parties must rely upon their own skill and judgement when making use of them. The author, publisher and contributors do not assume any liability to anyone for any loss or damage caused by any error or omission in the work, whether such an error or omission is the result of negligence or any other cause. Where reference is made to legislation it is not to be considered as legal advice. Any and all such liability is disclaimed.

ISBN 978-1-84919-287-3

Typeset in the UK by Phoenix Photosetting, Chatham, Kent
Printed in the UK by Polestar Wheatons, Exeter

Contents

Cooperating organisations	6
Preface	7
Foreword	9
Section 1 Introduction	**11**
1.1 Scope	11
1.2 Building Regulations	12
1.3 Basic information required	14
Section 2 The electrical supply	**15**
2.1 General layout of equipment	15
2.2 Function of components	17
2.3 Separation of gas installation pipework from other services	18
2.4 Portable generators	19
Section 3 Protection	**23**
3.1 Types of protective device	23
3.2 Overload protection	23
3.3 Fault current protection	23
3.4 Protection against electric shock	24
3.5 Automatic disconnection	25
3.6 Residual current devices (RCDs)	26
3.7 Surge protective devices (SPDs)	31
Section 4 Earthing and bonding	**39**
4.1 Protective earthing	39
4.2 Legal requirements	39
4.3 Main protective bonding of metallic services	39
4.4 Earthing conductor and main protective bonding conductor cross-sectional areas	40
4.5 Main protective bonding of plastic services	41
4.6 Supplementary equipotential bonding	42
4.7 Additional protection – supplementary equipotential bonding	42
4.8 Supplementary bonding of plastic pipe installations	43
4.9 Earth electrode	43
4.10 Types of earth electrode	43
4.11 Typical earthing arrangements for various types of earthing system	44

Section 5	**Isolation and switching**	**45**
5.1	Isolation	45
5.2	Switching off for mechanical maintenance	46
5.3	Emergency switching	46
5.4	Functional switching	47
5.5	Firefighter's switch	47
Section 6	**Labelling**	**49**
6.1	Additional protection	49
6.2	Retention of a dangerous electrical charge	49
6.3	Where the operator cannot observe the operation of switchgear and controlgear	49
6.4	Unexpected presence of nominal voltage exceeding 230 V	49
6.5	Connection of earthing and bonding conductors	50
6.6	Purpose of switchgear and controlgear	50
6.7	Identification of protective devices	50
6.8	Identification of isolators	50
6.9	Isolation requiring more than one device	50
6.10	Periodic inspection and testing	51
6.11	Diagrams	51
6.12	Residual current devices	51
6.13	Warning notice – non-standard colours	52
6.14	Warning notice – alternative supplies	52
6.15	Warning notice – high protective conductor current	53
6.16	Warning notice – photovoltaic systems	54
Section 7	**Final circuits**	**55**
7.1	Final circuits	55
7.2	Standard final circuits	68
7.3	Installation considerations	73
7.4	Proximity to electrical and other services	75
7.5	Earthing requirements for the installation of equipment having high protective conductor current	77
7.6	Electrical supplies to furniture	79
Section 8	**Locations containing a bath or shower**	**81**
8.1	Summary of requirements	81
8.2	Shower cubicle in a room used for other purposes	84
8.3	Underfloor heating systems	84
Section 9	**Inspection and testing**	**85**
9.1	Inspection and testing	85
9.2	Inspection	85
9.3	Testing	87
Section 10	**Guidance on initial testing of installations**	**89**
10.1	Safety and equipment	89
10.2	Sequence of tests	90
10.3	Test procedures	90
Section 11	**Operation of RCDs**	**105**
11.1	General test procedure	106
11.2	General-purpose RCCBs to BS 4293	106

11.3	General-purpose RCCBs to BS EN 61008 or RCBOs to BS EN 61009	106
11.4	RCD protected socket-outlets to BS 7288	106
11.5	Additional protection	106
11.6	Integral test device	107
11.7	Multipole RCDs	107

Appendix A	Maximum demand and diversity	109
Appendix B	Maximum permissible measured earth fault loop impedance	113
Appendix C	Selection of types of cable for particular uses and external influences	121
Appendix D	Methods of support for cables, conductors and wiring systems	127
Appendix E	Cable capacities of conduit and trunking	133
Appendix F	Current-carrying capacities and voltage drop for copper conductors	139

Appendix G	Certification and reporting	151
G1	Introduction	151
G2	Certification	151
G3	Reporting	152
G4	Introduction to Model Forms from BS 7671:2008(2011)	153

Appendix H	Standard circuit arrangements for household and similar installations	173
H1	Introduction	173
H2	Final circuits using socket-outlets complying with BS 1363-2 and fused connection units complying with BS 1363-4	173
H3	Radial final circuits using 16 A socket-outlets complying with BS EN 60309-2 (BS 4343)	176
H4	Cooker circuits in household and similar premises	176
H5	Water and space heating	177
H6	Height of switches, socket-outlets and controls	177
H7	Number of socket-outlets	178

Appendix I	Resistance of copper and aluminium conductors	181
Appendix J	Selection of devices for isolation and switching	185

Appendix K	Identification of conductors	187
K1	Introduction	187
K2	Addition or alteration to an existing installation	189
K3	Switch wires in a new installation or an addition or alteration to an existing installation	189
K4	Intermediate and two-way switch wires in a new installation or an addition or alteration to an existing installation	190
K5	Line conductors in a new installation or an addition or alteration to an existing installation	190
K6	Changes to cable core colour identification	190
K7	Addition or alteration to a d.c. installation	191

Index   193

Cooperating organisations

The IET acknowledges the invaluable contribution made by the following organisations in the preparation of this guide.

BEAMA Installation Ltd.
P. Sayer IEng MIET GCGI

British Cables Association
I. Collings BEng (Hons) CEng MIMechE
M. Gaucher
C.K. Reed IEng MIET

British Electrotechnical & Allied Manufacturers Association Ltd
P. Still MIEE

British Gas/Micropower Council
P. Gibson

British Standards Institution
P. Calver – Chairman BSI FW0/3
A.S. Khan MEng(Hons) MIET MIEEE – PEL 37/1, GEL 81

City & Guilds of London Institute
H.R. Lovegrove IEng FIET

Department for Communities and Local Government
K. Bromley

Electrical Contractors' Association
C. Flynn IEng MIET (Elec) CGI

Electrical Contractors' Association of Scotland t/a SELECT
R. Cairney IEng MIET
M.M. Duncan IEng MIET MILP

ENA
T. Haggis

ESSA
P. Yates MSc MIEE

ERA Technology Ltd
M.W. Coates BEng

Electrical Safety Council
G. Gundry

Health and Safety Executive
K. Morton BSc CEng FIET

IHEEM
Eur Ing P. Harris BEng(Hons) FIHEEM MIEE MCIBSE

Institution of Engineering and Technology
G.D. Cronshaw IEng FIET
P.E. Donnachie BSc CEng FIET
Eur Ing D. Locke BEng(Hons) CEng MIET MIEEE
Eur Ing L. Markwell MSc BSc(Hons) CEng MIEE MCIBSE LCGI LCC
I.M. Reeve BTech CEng MIEE

National Inspection Council for Electrical Installation Contracting
J.M. Maltby-Smith BSc(Hons) PG Dip Cert Ed IEng MIET

NAPIT
W.R. Allan BEng(Hons)

Safety Assessment Federation
I. Trueman CEng MSOE MBES MIET

Society of Electrical and Mechanical Engineers serving Local Government
C.J. Tanswell CEng MIET MCIBSE

UHMA
Dr S. Newberry

Revised, compiled and edited
M. Coles BEng(Hons) MIEE, The Institution of Engineering and Technology, 2011

Preface

The *On-Site Guide* is one of a number of publications prepared by the IET to provide guidance on certain aspects of BS 7671:2008(2011) *Requirements for Electrical Installations (IET Wiring Regulations, 17th Edition, incorporating Amendment No. 1)*. BS 7671 is a joint publication of the British Standards Institution and the Institution of Engineering and Technology.

110.1 The scope generally follows that of BS 7671. The Guide includes material not included in BS 7671, provides background to the intentions of BS 7671 and gives other sources of information, however, it does not ensure compliance with BS 7671. It is a simple guide to the requirements of BS 7671 and electrical installers should always consult BS 7671 to satisfy themselves of compliance.

It is expected that persons carrying out work in accordance with this guide will be competent to do so.

HSR25, Electrical installations in the United Kingdom which comply with the *IET Wiring*
EWR *Regulations*, BS 7671, must comply with all relevant statutory regulations, such as the
Regulation Electricity at Work Regulations 1989, the Building Regulations and, where relevant, the
16 Electricity Safety, Quality and Continuity Regulations 2002 and Amendment 2006.

114.1 It cannot be guaranteed that BS 7671 complies with all relevant statutory regulations.
115.1 It is, therefore, essential to establish which statutory and other appropriate regulations apply and to install accordingly; for example, an installation in licensed premises may have requirements which differ from or are additional to BS 7671 and these will take precedence.

Foreword

Part 1 This Guide is concerned with limited application of BS 7671 in accordance with paragraph 1.1: Scope.

BS 7671 and the On-Site Guide are not design guides.

It is essential to prepare a design and/or schedule of the work to be done prior to commencement or alteration of an electrical installation and to provide all necessary information and operating instructions of any equipment supplied to the user on completion.

Any specification should set out the detailed design and provide sufficient information to enable competent persons to carry out the installation and commissioning.

The specification must provide for all the commissioning procedures that will be required and for the production of any operation and maintenance manual and building logbook.

The persons or organisations who may be concerned in the preparation of the specification include:

- the Designer(s)
- the Installer(s)
- the Electricity Distributor
- the Installation Owner and/or User
- the Architect
- the Local Building Control Authority/Standards division or Approved Inspector
- the Fire Prevention Officer
- the CDM Coordinator
- all Regulatory Authorities
- any Licensing Authority
- the Health and Safety Executive.

In producing the specification, advice should be sought from the installation owner and/or user as to the intended use. Often, such as in a speculative building, the detailed intended use is unknown. In those circumstances the specification and/or the operation and maintenance manual and building logbook must set out the basis of use for which the installation is suitable.

Precise details of each item of equipment should be obtained from the manufacturer and/or supplier and compliance with appropriate standards confirmed.

The operation and maintenance manual must include a description of how the installed system is to operate and must include all commissioning records. The manual should also include manufacturers' technical data for all items of switchgear, luminaires, accessories, etc. and any special instructions that may be needed.

Part L 2010 of the Building Regulations of England and Wales requires that building owners or operators are provided with summary information relating to a new or refurbished building which includes building services information and the maintenance requirements in a building logbook. Information on how to develop and assemble a building logbook can be obtained from CIBSE:

Tel.: 020 8772 3618
Website: www.cibse.org
Address: CIBSE
222 Balham High Road
London
SW12 9BS

The Health and Safety at Work etc. Act 1974 Section 6 and the Construction (Design and Management) Regulations 2007 are concerned with the provision of information. Guidance on the preparation of technical manuals is given in BS 4884 series *Technical manuals* and BS 4940 series *Technical information on construction products and services*. The size and complexity of the installation will dictate the nature and extent of the manual.

Introduction 1

1.1 Scope

This Guide is for installers (for simplicity, the term *installer* has been used for electricians and electrical installers). It covers the following installations:

    a domestic and similar installations, including off-peak supplies, supplies to associated garages, outbuildings and the like
    b small industrial and commercial single- and three-phase installations.

Part 7 **NOTE:** Special Installations or Locations (Part 7 of BS 7671) are generally excluded from this Guide, as are installations for electric vehicle charging equipment. Advice, however, is given on installations in locations containing a bath or shower and underfloor heating installations.

This Guide is restricted to installations:

313.1
    i at a supply frequency of 50 hertz
    ii at a nominal voltage of 230 V a.c. single-phase or 400/230 V a.c. three-phase
    iii supplied through a distributor's cut-out having a fuse or fuses rated at 100 A or less to one of the following standards:
        – BS 88-2
        – BS 88-3
        – BS 88-6
        – BS 1361 Type II

NOTE: BS 1361 was withdrawn in March 2010 and replaced by BS 88-3; BS 88-6 was withdrawn in March 2010 and replaced by BS 88-2 but fuses complying with these withdrawn standards will be found in existing installations for many years to come.

    iv typical maximum values of earth fault loop impedance, Z_e, for TN earthing arrangements outside the consumer's installation commonly quoted by distributors are as follows:
        ▶ TN-C-S arrangement - 0.35 Ω, see Figure 2.1(i)
        ▶ TN-S arrangement - 0.8 Ω, see Figure 2.1(ii)

Table 41.5
542.2.4
  For a TT arrangement, 21 Ω is the usual stated maximum resistance of the distributor's earth electrode at the supply transformer. The resistance of the consumer's installation earth electrode should be as low as practicable and a value exceeding 200 Ω may not be stable.

This Guide also contains information which may be required in general installation work, for example, conduit and trunking capacities, bending radii of cables, etc.

The Guide introduces the use of standard circuits, which are discussed in Section 7, however, because of simplification, this Guide may not give the most economical result.

This Guide is not a replacement for BS 7671 which should always be consulted.

Defined terms according to Part 2 of BS 7671 are used.

In compliance with the definitions of BS 7671, throughout this Guide the term *line conductor* is used instead of *phase conductor* and *live part* is used to refer to a conductor or conductive part intended to be energised in normal use, including a neutral conductor.

The terminals of electrical equipment are identified by the letters L, N and E (or PE).

Further information is available in the series of Guidance Notes published by the IET.

NOTE: For clarification:

- the *distributor* of electricity is deemed to be the organisation owning the electrical supply equipment, and
- the *supplier* of electricity is the organisation from whom electricity is purchased.

1.2 Building Regulations

Refer to the IET publication *Electrician's Guide to the Building Regulations* for more in-depth guidance on electrical installations in dwellings.

1.2.1 The Building Regulations of England and Wales

Persons carrying out electrical work in dwellings must comply with the Building Regulations of England and Wales, in particular Part P (Electrical safety – dwellings).

Persons responsible for work within the scope of Part P of the Building Regulations may also be responsible for ensuring compliance with other Parts of the Building Regulations, where relevant, particularly if there are no other parties involved with the work. Building Regulations requirements relevant to installers carrying out electrical work include:

Part A	Structure: depth of chases in walls and size of holes and notches in floor and roof joists;
Part B	Fire safety: fire safety of certain electrical installations; provision of fire alarm and fire detection systems; fire resistance of penetrations through floors and walls;
Part C	Site preparation and resistance to moisture: moisture resistance of cable penetrations through external walls;
Part D	Toxic substances;
Part E	Resistance to the passage of sound: penetrations through floors and walls;
Part F	Ventilation: ventilation rates for dwellings;
Part G	Sanitation, hot water safety and water efficiency;

Part J   Heat producing appliances;
Part K   Protection from falling;
Part L   Conservation of fuel and power: energy efficient lighting;
Part M   Access to and use of buildings: heights of switches, socket-outlets and consumer units;
Part P   Electrical safety – dwellings.

NOTE: Guidance is available for each part of the Building Regulations in the form of Approved Documents which can be freely downloaded from the Department for Communities and Local Government (DCLG) website: www.planningportal.gov.uk

1.2.2 The Building (Scotland) Regulations 2004

The detailed requirements are given in the Technical Standards for compliance with the Building (Scotland) Regulations.

Guidance on how to achieve compliance with these Standards is given in two Scottish Building Standards Technical Handbooks – Domestic and Non-domestic.

These handbooks contain recommendations for electrical installations including the following:

- compliance with BS 7671
- minimum number of socket-outlets in dwellings
- minimum number of lighting points in dwellings
- minimum illumination levels in common areas of domestic buildings, for example, blocks of flats
- a range of mounting heights of switches and socket-outlets, etc
- separate switching for concealed socket-outlets, for example, behind white goods in kitchens
- conservation of fuel and power in buildings.

With regard to electrical installations in Scotland, the requirements of the above are deemed to be satisfied by complying with BS 7671.

NOTE: The handbooks may be obtained from the Building Standards Division of the Scottish Government from website:
www.scotland.gov.uk/Topics/Built-Environment/Building/Building-standards/publications/pubtech

1.2.3 The Building Regulations of Northern Ireland

The Building Regulations (Northern Ireland) 2000 (as amended) apply.

NOTE: Information can be obtained from the website: www.buildingcontrol-ni.com

1.3 Basic information required

313.1

Before starting work on an installation which requires a new electrical supply, the installer should establish the following information with the local distributor:

- **i** the number of live conductors required by the design
- **ii** the distributor's requirement for cross-sectional area and length of meter tails
- **iii** the maximum prospective fault current (Ipf) at the supply terminals
- **iv** the typical maximum earth fault loop impedance (Ze) of the earth fault path outside the consumer's installation
- **v** the type and rating of the distributor's fusible cut-out or protective device

544.1

- **vi** the distributor's requirement regarding the size of main protective bonding conductors
- **vii** the conductor arrangement and system earthing

312

- **viii** the arrangements for the incoming cable and metering.

132.16 For additions and alterations to existing installations, installers should satisfy themselves as to the suitability of the supply, the distributor's equipment and the earthing arrangements.

The electrical supply 2

2.1 General layout of equipment

The general layout of the equipment at the service position is shown in Figures 2.1(i) to 2.1(iii) including typical protective conductor cross-sectional areas.

▼ **Figure 2.1(i)** TN-C-S (PME) earthing arrangement

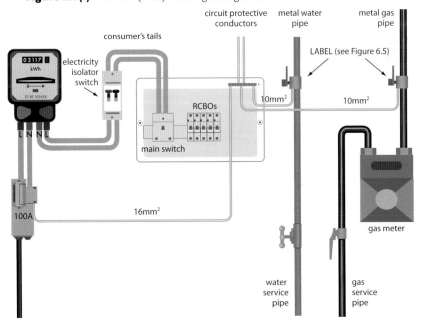

NOTE: An electricity isolator switch may not always be installed by the distributor.

▼ **Figure 2.1(ii)** TN-S earthing arrangement (cable sheath earth)

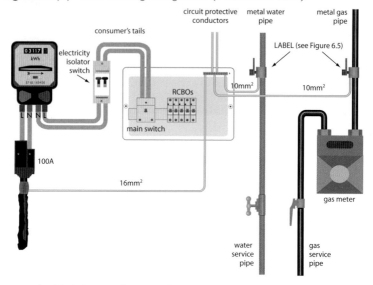

NOTE: An electricity isolator switch may not always be installed by the distributor.

▼ **Figure 2.1(iii)** TT earthing arrangement (no distributor's earth)

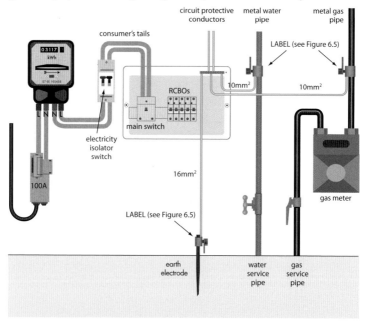

NOTE 1: An electricity isolator switch may not always be installed by the distributor.

542.3.1 **NOTE 2**: See Table 4.4(ii) for further information regarding the sizing of the earthing conductor for a TT earthing arrangement.

2.2 Function of components

2.2.1 Distributor's cut-out

This will be sealed to prevent the fuse being withdrawn by unauthorised persons. When the meter tails and consumer unit are installed in accordance with the requirements of the distributor, the cut-out may be assumed to provide fault current protection up to the consumer's main switch.

As the cut-out is the property of the distributor, installers must not cut seals and withdraw cut-out fuses without permission. When removal of the cut-out for isolation is required, the supplier of electricity should be contacted to arrange disconnection and subsequent reconnection.

NOTE: The supplier of electricity may not be the same organisation as the distributor.

2.2.2 Electricity meter

The terminals will be sealed by the meter owner to prevent interference by unauthorised persons.

2.2.3 Meter tails

521.10.1 Meter tails are part of the consumer's installation and should be insulated and sheathed or insulated and enclosed within containment, for example, conduit or trunking. Meter tails are provided by the installer and are the responsibility of the owner of the electrical installation.

514.3 Polarity should be indicated by the colour of the insulation and the minimum cable size should be 25 mm^2. The distributor may specify the maximum length of meter tails between the meter and the consumer unit in addition to the minimum cross-sectional area (see 1.3). In some cases, the distributor may require an electricity isolator switch (see 2.2.4).

434.3(iv) Where the meter tails are protected against fault current by the distributor's cut-out, the method of installation, maximum length and minimum cross-sectional area must comply with the requirements of the distributor.

522.6.101 Where meter tails are buried in walls, further protection is required (see 7.3.2).

2.2.4 Electricity isolator switch

Suppliers may provide and install an electricity isolator switch between the meter and the consumer unit, labelled as *Electricity isolator switch* in Figures 2.1(i) to 2.1(iii). This double-pole switch permits the supply to the installation to be interrupted without withdrawing the distributor's cut-out fuse.

2

2.2.5 Consumer's controlgear

530.3.4 A consumer unit (to BS EN 60439-3 Annex ZA) is for use on single-phase installations up to 100 A and may include the following components:

- a double-pole isolator
- fuses, circuit-breakers or RCBOs for protection against overload and fault currents
- RCDs for additional protection against electric shock
- RCDs for fault protection.

Alternatively, a separate main switch and distribution board may be provided.

2.3 Separation of gas installation pipework from other services

Gas installation pipes must be spaced:

a at least 150 mm away from electricity meters, controls, electrical switches or socket-outlets, distribution boards or consumer units;
b at least 25 mm away from electricity supply and distribution cables.

528.3.4 (The cited distances are quoted within BS 6891:2005+A2:2008 *Installation of low*
Note *pressure gas pipework in domestic premises*, clause 8.16.2.)

▼ **Figure 2.3** Separation from gas pipes and gas metering equipment

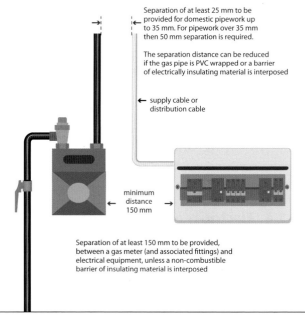

2.4 Portable generators

551.4.4 It is recognised that generators will be used occasionally as a temporary or short-term means of supplying electricity, for example:

- ▶ use on a construction site
- ▶ used to supply stalls on street markets
- ▶ external gathering or function attended by the general public, such as a country show.

Temporary generators can be divided into two classes, i.e. portable and mobile:

- ▶ portable generators with an electrical output rating of up to 10 kVA are used for small-scale work for short-term use, i.e. less than one day, and
- ▶ mobile generators are those used for longer periods and can be in excess of 10 kVA output.

This guide considers three scenarios relating to the use of portable generators; see 2.4.1 to 2.4.3.

551 For information relating to the permanent use of generators see IET Guidance Notes 5 and 7 and Section 551 of BS 7671:2008(2011).

Where generators are used to supply concession vehicles, such as burger vans, see Section 717 Mobile and Transportable Units of BS 7671:2008(2011) and IET Guidance Note 7.

2.4.1 Portable generator used with a floating earth

Small portable generators, ranging in output from 0.3 kVA to 10 kVA single-phase often have a floating earth, i.e. there is no connection between the chassis and/or earth connection of the socket-outlet of the unit to the neutral of the generator winding. The ends of the generator winding are brought out to one or more three-pin socket-outlets which should conform to BS EN 60309-2. The earth socket-tube of the socket-outlet(s) 551.4.4 are usually connected internally to the frame of the generator only; see Figure 2.4.1.

This arrangement is a form of *electrical separation*, where basic protection is provided by basic insulation of live parts and fault protection is provided by simple separation of the separated circuit from other circuits and from Earth. The requirements for electrical separation can be 413 found in Section 413 of BS 7671 where one item of equipment is supplied and Regulation 418.3 418.3 where more than one item of equipment is supplied by the separated circuit.

It is extremely important to note that a portable generator used with floating earth should only be used to supply equipment in the following permutations:

- ▶ one or more items of Class II equipment
- ▶ one item of Class I equipment
- ▶ one or more items of Class II and one item of Class I equipment.

The use of only Class II equipment, however, is preferable.

More than one item of Class I equipment should not be used simultaneously as faults can be presented as voltages and operatives can provide a path for current flowing between exposed-conductive-parts of faulty electrical equipment.

▼ **Figure 2.4.1**   Portable generator used with a floating earth

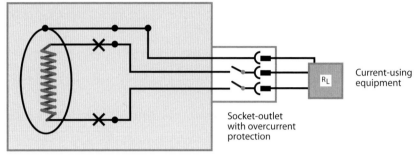

Generator

2.4.2 Portable generator used without reference to the general mass of the earth

551.4.4   Where more than one item of Class I equipment is to be supplied by a single-phase portable generator, it is important to ensure that the earth connections of the socket-outlets at the generator are connected to the neutral of the generator winding in addition to the chassis or frame of the generator. See Figure 2.4.2.

Such a configuration will provide a return path for any fault current caused by contact between live parts and exposed-conductive-parts of the connected equipment. If this method of supply is used, extreme care should be taken to ensure that there is no intended or casual interconnection with any other electrical system, such as extraneous-conductive-parts or exposed-conductive-parts from other electrical systems.

RCD protection at 30 mA is required for all circuits supplied in this manner.

▼ **Figure 2.4.2**   Generator supplying more than one item of equipment

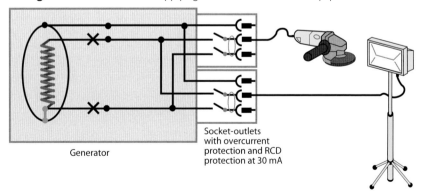

2.4.3 Portable generator referenced to the general mass of the earth

BS 7430: 1998
Where there are extraneous-conductive-parts or exposed-conductive-parts from other electrical systems present, generator reference earthing, by means of an earth electrode to the general mass of the earth, should be installed. See Figure 2.4.3(i).

Note that this does not create a TT supply arrangement; the supply will be TN-S in form from the generator, the neutral or star point being referenced to the general mass of the earth.

Where an earth electrode is supplied it will need to be tested by the standard method using a proprietary earth electrode resistance tester; see 10.3.5.2.

Note that an earth fault loop impedance tester cannot be used for this test as the earth electrode is not used as a means of earthing, it is used to reference the portable generator to the general mass of the earth and does not form part of the earth loop.

As the earth electrode is used for referencing and not as a means of earthing, its resistance should, ideally, be less than 200 Ω.

Table 54.1
543.3.1
If buried, generator reference earthing and/or bonding conductors should be sized in accordance with Table 54.1 and suitably protected in accordance with Regulation 543.3.1. For example, a 16 mm² conductor would generally be adequate for short-term use where no mechanical protection is provided.

▼ **Figure 2.4.3(i)** Generator reference earthing - using earth electrode

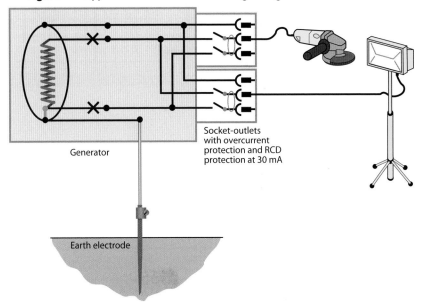

2

Where restrictions, such as concreted/paved areas or the portable generator is being used some distance above ground level, make it impossible to install an earth electrode, simultaneously accessible metal parts, i.e. accessible extraneous-conductive-parts and/or exposed-conductive-parts from other electrical systems, may be bonded to the main earthing terminal of the generator. See Figure 2.4.3(ii).

544.1.1 Where separate accessible extraneous-conductive-parts and/or exposed-conductive-parts from other electrical systems are connected together, protective conductors can be sized in accordance with Regulation 544.1.1. For example, a 16mm^2 conductor would generally be adequate for short-term use where no mechanical protection is provided.

▼ **Figure 2.4.3(ii)** Generator reference earthing - connection of extraneous- and/or exposed-conductive-parts where the installation of an earth electrode is not possible

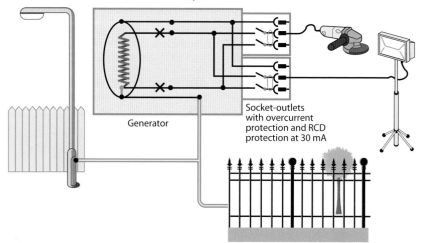

Protection 3

3.1 Types of protective device

The consumer unit (or distribution board) contains devices for the protection of distribution circuits and final circuits against:

433
434
434

   i   overload
   ii short-circuit
   iii earth fault.

Functions i and ii are carried out usually by one device, i.e. a fuse or circuit-breaker.

434
411

Function iii may be carried out by the fuse or circuit-breaker provided for functions i and ii or by an RCD.

An RCBO, being a unit with a combined circuit-breaker and RCD, will carry out functions i, ii and iii.

Appx 3
533.1

3.2 Overload protection

Overload protection will be provided by the use of any of the following devices:

- ▶ fuses to BS 88-2, BS 88-3, BS 88-6, BS 1361 and BS 3036
- ▶ miniature circuit-breakers to BS 3871-1 types 1, 2 and 3
- ▶ circuit-breakers to BS EN 60898 types B, C and D, and
- ▶ residual current circuit-breakers with integral overcurrent protection (RCBOs) to BS EN 61009-1 and IEC 62325.

3.3 Fault current protection

When a consumer unit to BS EN 60439-3 or BS 5486:Part 13 or a fuseboard having fuselinks to BS 88-2 or BS 88-6 or BS 1361 is used, then fault current protection will be given by the overload protective device.

For other protective devices the breaking capacity must be adequate for the prospective fault current at that point.

3.4 Protection against electric shock

3.4.1 Automatic disconnection of supply

Automatic disconnection of supply (ADS) is the most the common method of protection against electric shock. There are two elements to automatic disconnection of supply, *basic protection* and *fault protection*.

3.4.1.1 Basic protection

Basic protection is the physical barrier between persons/livestock and a live part. Examples of basic protection are:

- ▶ electrical insulation
- ▶ enclosures and barriers.

It follows that single-core non-sheathed insulated conductors must be protected by conduit or trunking and be terminated within a suitable enclosure.

A 30 mA RCD may be provided to give additional protection against contact with live parts but must not be used as primary protection.

3.4.1.2 Fault protection

Fault protection comprises:

- ▶ protective earthing,
- ▶ protective equipotential bonding, and
- ▶ automatic disconnection in case of a fault.

Fault protection is provided by limiting the magnitude and duration of voltages that may appear under earth fault conditions between simultaneously accessible exposed-conductive-parts of equipment and between them and extraneous-conductive-parts or earth.

3.4.2 Other methods of protection against electric shock

In addition to automatic disconnection of supply, BS 7671 recognises other methods of protection against electric shock.

3.4.3 SELV and PELV

SELV

Separated extra-low voltage (SELV) systems:

- ▶ are supplied from isolated safety sources such as a safety isolating transformer to BS EN 61558-2-6
- ▶ have no live part connected to earth or the protective conductor of another system
- ▶ have basic insulation from other SELV and PELV circuits
- ▶ have double or reinforced insulation or basic insulation plus earthed metallic screening from LV circuits

- have no exposed-conductive-parts connected to earth or to exposed-conductive-parts or protective conductors of another circuit.

PELV

Protective extra-low voltage (PELV) systems must meet all the requirements for SELV, except that the circuits are not electrically separated from earth.

For SELV and PELV systems basic protection need not be provided if voltages do not exceed those given in Table 3.4.3.

▼ **Table 3.4.3** SELV and PELV basic protection voltage limits

Location	SELV and PELV
Dry areas	25 V a.c. or 60 V d.c
Immersed equipment	Further protection required at all voltages
Locations containing a bath or shower, swimming pools, saunas	Further protection required at all voltages
Other areas	12 V a.c. or 30 V d.c.

3.5 Automatic disconnection

3.5.1 Standard circuits

For the standard final circuits given in Section 7 of this Guide, the correct disconnection time is obtained for the protective devices by limiting the maximum circuit lengths.

3.5.2 Disconnection times – TN circuits

A disconnection time of not more than 0.4 s is required for final circuits with a rating (I_n) not exceeding 32 A.

A disconnection time of not more than 5 s is permitted for:

- final circuits exceeding 32 A, and
- distribution circuits.

3.5.3 Disconnection times – TT circuits

The required disconnection times for installations forming part of a TT system can, except in the most exceptional circumstances outside the scope of this guide, only be achieved by protecting every circuit with an RCD, hence, a time of not more than 0.2 s is required for final circuits with a rating (I_n) not exceeding 32 A.

A disconnection time of not more than 1 s is permitted for:

- final circuits exceeding 32 A, and
- distribution circuits.

3

3.6 Residual current devices (RCDs)

RCD is the generic term for a device that operates when the residual current in the circuit reaches a predetermined value. The RCD is, therefore, the main component in an RCCB (residual current operated circuit-breaker without integral overcurrent protection) or one of the functions of an RCBO (residual current operated circuit-breaker with integral overcurrent protection).

3.6.1 Protection by RCDs

RCDs are required:

411.5	i	where the earth fault loop impedance is too high to provide the required disconnection, for example, where the distributor does not provide a connection to the means of earthing – TT earthing arrangement
411.3.3(i)	ii	for socket-outlets where used by ordinary persons for general use
701.411.3.3	iii	for all circuits of locations containing a bath or shower
411.3.3(ii)	iv	for circuits supplying mobile equipment not exceeding 32 A for use outdoors
522.6.101	v	for cables without earthed metallic covering installed in walls or partitions at a depth of less than 50 mm and not protected by earthed steel conduit or similar
522.6.102		
522.6.103	vi	for cables without earthed metallic covering installed in walls or partitions with metal parts (not including screws or nails) and not protected by earthed steel conduit or the like.

3.6.2 Omission of RCD protection

3.6.2.1 Specific cases

RCD protection can be omitted in the following circumstances:

411.3.3(b)	i	specific labelled socket-outlets, for example, a socket-outlet for a freezer. However, the circuit cables must not require RCD protection as per v and vi in clause 3.6.1, that is, circuit cables must be enclosed in earthed steel conduit or have an earthed metal sheath or be at a depth of at least 50 mm in a wall or partition without metal parts
411.3.3(a)	ii	socket-outlet circuits in situations where the use of equipment and work on the building fabric and electrical installation is controlled by skilled or instructed persons, for example, in some industrial and commercial locations; see 3.6.2.2.
411.5		Cables installed on the surface do not specifically require RCD protection, however, RCD protection may be required for other reasons, for example, for fault protection, where the earth fault loop impedance is such that the disconnection time for an overcurrent device cannot be met.
411.3.3(b)		It is expected that all socket-outlets in a dwelling will have RCD protection at 30 mA, however, the exception of Regulation 411.3.3 can be applied in certain cases.

3.6.2.2 Installations under the control of skilled or instructed persons

411.3.3(a)
522.6.102
522.6.103

BS 7671:2008(2011) permits RCDs, where usually provided for additional protection, to be omitted where the installation is under the control of a skilled or instructed person.

411.3.3(i)
415.1.1

The decision as to which socket-outlets or circuits do not require additional protection by RCDs should be taken by the designer of the electrical installation and only after consultation with an appropriate person in the client's organisation. An appropriate person would be one who is able to ensure that the socket-outlets or circuits in question are, and will remain, under the supervision of skilled or instructed persons.

Wherever a designer so chooses to omit RCD protection, traceable confirmation must be obtained from the client to identify the reason for the omission and such confirmation must be included within the documentation handed over to the client upon completion of the work.

Where no such confirmation can be obtained, RCD protection should not be omitted.

3.6.3 Applications of RCDs

314

Installations are required to be divided into circuits to avoid hazards and minimize inconvenience in the event of a fault and to take account of danger that might arise from the failure of a single circuit, such as a lighting circuit.

The following scenarios show different methods of providing RCD protection within installations. Note that, for clarity, earthing and bonding connections are not shown.

a TN conduit installations

Where cables in walls or partitions have an earthed metallic covering or are installed in steel conduit or similar, 30 mA RCD protection is still required in the following cases:

- ▶ circuits of locations containing a bath or shower
- ▶ protection at socket-outlets not exceeding 20 A
- ▶ mobile equipment not exceeding 32 A for use outdoors
- ▶ the arrangement in Figure 3.6.3(i).

▼ **Figure 3.6.3(i)** Typical split consumer unit with one 30 mA RCD, suitable for TN installations with cables in walls or partitions having an earthed metallic covering or enclosed in earthed steel conduit or the like

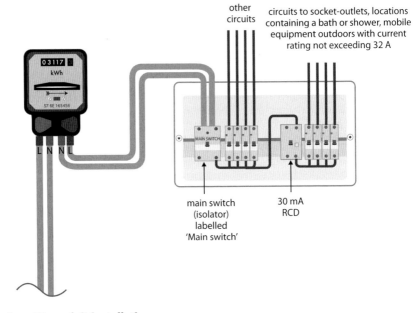

b   TT conduit installations

For installations forming part of a TT system, all circuits must be RCD protected. If cables in walls or partitions have an earthed metallic covering or are installed in earthed steel conduit, 30 mA RCDs will be required for:

- ▶ circuits of locations containing a bath or shower
- ▶ circuits with socket-outlets not exceeding 20 A
- ▶ mobile equipment not exceeding 32 A for use outdoors.

The remainder of the installation would require protection by a 100 mA RCD (see Figure 3.6.3(ii)).

▼ **Figure 3.6.3(ii)** Typical split consumer unit with time-delayed RCD as main switch, suitable for TT and TN installations with cables in walls or partitions having an earthed metallic covering or enclosed in earthed steel conduit or the like

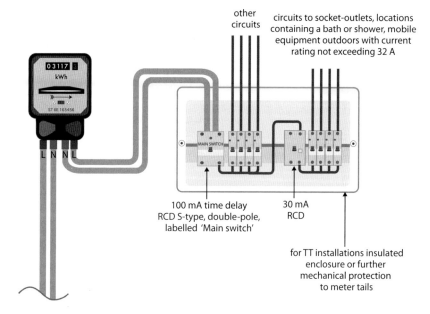

For installations forming part of a TT system with cables installed in walls or partitions having no earthed metallic covering or not installed in earthed conduit or the like, protection by 30 mA RCDs will be required for all circuits, see Figures 3.6.3(iii) and 3.6.3(iv).

The enclosures of RCDs or consumer units incorporating RCDs in TT installations should have an all-insulated or Class II construction or additional precautions, as may be recommended by the manufacturer, need to be taken to prevent faults to earth on the supply side of the 100 mA RCD.

c   RCBOs

The use of RCBOs will minimize inconvenience in the event of a fault and is applicable to all systems. See Figure 3.6.3(iii).

Such a consumer unit arrangement also easily allows individual circuits, such as to specifically labelled socket-outlets or fire alarms, to be protected by a circuit-breaker without RCD protection. Such circuits will usually need to be installed in earthed metal conduit or wired with earthed metal-sheathed cables.

▼ **Figure 3.6.3(iii)**   Consumer unit with RCBOs, suitable for all installations (TN and TT)

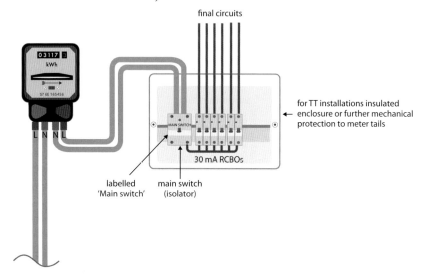

d   Split board with two 30 mA RCDs

The division of an installation into two parts with separate 30 mA RCDs will ensure that part of the installation will remain on supply in the event of a fault, see Figure 3.6.3(iv).

▼ **Figure 3.6.3(iv)**   Split consumer unit with separate main switch and two 30 mA RCDs

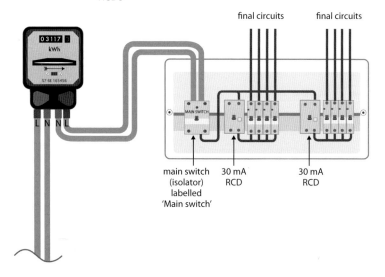

e   Three-way split board with two 30 mA RCDs

The three-way division of an installation can provide ways unprotected by RCDs for, say, fire systems and for two separate 30 mA RCDs to ensure that part of the installation will remain on supply in the event of a fault. Unprotected circuits will usually need to be installed in earthed metal conduit or wired with earthed metal-sheathed cables, see Figure 3.6.3(v).

▼ **Figure 3.6.3(v)**   Three-way split consumer unit with separate main switch, two 30 mA RCDs and circuits without RCD protection

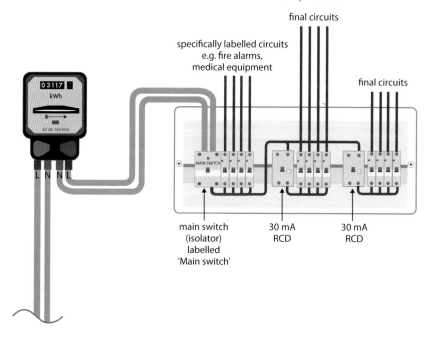

3.7   Surge protective devices (SPDs)

3.7.1   Overview

Electrical installations and connected equipment can be severely affected by lightning activity during thunderstorms or from electrical switching events.

For more information, see IET Guidance Note 1.

Damage can occur when the surge or transient overvoltage, as the result of lightning or electrical switching, exceeds the impulse withstand voltage rating of electrical equipment – the levels of which are defined in Table 44.3.

Surges from electrical switching events are created when large inductive loads, such as motors or air conditioning units, switch off and release stored energy which dissipates as a transient overvoltage. Switching surges are, in general, not as severe as lightning surges but are more repetitive and can reduce equipment lifespan.

Overvoltages of atmospheric origin, in particular, can present a risk of fire and electric shock owing to a dangerous flashover.

- ▶ Section 443 of BS 7671:2008(2011) has requirements for the protection of persons, livestock and property from injury and damage as a consequence of overvoltage
- ▶ Section 534 has requirements for the selection and installation of surge protective devices.

NOTE 1: Section 534 applies to a.c. power circuits only. When the need for power SPDs is identified, additional SPDs on other services such as telecommunications lines and equipment is also recommended. See BS EN 62305 and BS EN 61643.

NOTE 2: Some electronic equipment may have protection levels lower than Category I of Table 44.3.

3.7.2 Arrangements for protection against overvoltages

Protection according to Section 443 can only be achieved if transient overvoltages are limited to values lower than those given in Table 44.3, requiring the correct selection and installation of suitable SPDs.

3.7.2.1 Where SPD protection may not be required

Protection against overvoltages of atmospheric origin is not required in the following circumstances but, in each case, the impulse withstand voltage of equipment must meet the requirements of Table 44.3 of BS 7671:2008(2011):

- ▶ the installation is supplied by a completely buried low voltage system and does not include overhead lines
- ▶ installations which include overhead lines but where the consequential losses are tolerable, e.g. typical urban dwelling, storage unit or farm building.

If there are risks of direct strikes to or near the structure or to the low voltage distribution line, overvoltage protection by SPDs is required in accordance with BS EN 62305 *Protection against lightning*.

3.7.2.2 Where SPD protection is required

Surge protective devices should be considered in the following circumstances:

- ▶ the low voltage supply to the installation, at some point, is provided by bare overhead conductors at risk of direct lightning strike
- ▶ the building requires or already has a lightning protection system (LPS)
- ▶ the risk of loss of any part of the installation or equipment due to damage caused by any transient overvoltages (including switching transients) is not acceptable.

The flow chart in Figure 3.7.2.2 will aid the decision-making process for electrical installations within the scope of this Guide. See IET Guidance Note 1 for more information.

▼ **Figure 3.7.2.2** SPD decision flow chart for installations within the scope of this Guide

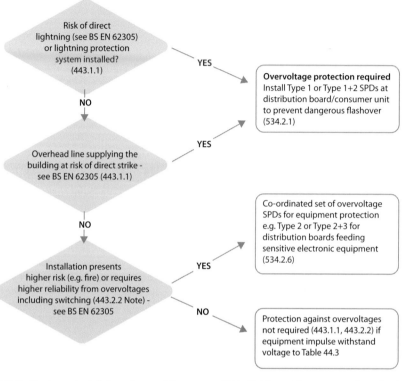

NOTE: For larger installations beyond the scope of this Guide, a risk assessment method used to evaluate the need for SPDs is given in Section 443 of BS 7671:2008(2011).

3.7.3 Types of SPD protection

For the protection of a.c. power circuits, SPDs are allocated a type number:

- ▶ Type 1 SPDs are only used where there is a risk of direct lightning current and, typically, are installed at the origin of the installation
- ▶ Type 2 SPDs are used at distribution boards
- ▶ Type 3 SPDs are used near terminal equipment.

See also Table 3.7.3.

Combined Type SPDs are classified with more than one Type, e.g. Type 1 & 2, Type 2 & 3, and can provide both lightning current with overvoltage protection in addition to protection between all conductor combinations (or modes of protection) within a single unit. Combined Type SPDs provide high surge current handling combined with better overvoltage protection levels (Up) – the latter being a performance parameter of an SPD.

▼ **Table 3.7.3** Types of SPD protection

Type	Name	Location	CSA conductor	Hazard
1	Equipotential bonding or lightning protection/ current SPD	Origin of the installation	16 mm² minimum – length of tails – ideally <0.5 m but no longer than 1 m	Protect against flashover from direct lightning strikes to structure or to LV overhead supply
2	Overvoltage SPD	Distribution board/ consumer unit	4 mm² or equal to CSA of circuit conductors	Protect against overvoltages which can overstress the electrical installation
3	Overvoltage SPD	Terminal equipment	2.5 mm² or equal to CSA of circuit conductors	Protect against overvoltages and high currents on items of equipment

3.7.4 Coordination and selection of surge protection

534.2.3.6 Where a number of SPDs are required to operate in conjunction with each other they must be coordinated to ensure the correct type of protection is installed where required; see Figure 3.7.4.

SPD protection should be coordinated as follows:

- ▶ choose the correct type of SPD for the installation and site in the correct location
- ▶ refer to Tables 44.3 and 44.4 of BS 7671 (impulse withstand voltage)

534.2.3.1.1
- ▶ choose SPDs with a protection level (Up) sufficiently lower than the impulse withstand voltage or lower than the impulse immunity of the equipment to be protected
- ▶ choose SPDs of the same make or manufacture.

NOTE: Coordinated SPDs must be of the same make or manufacture unless the designer is satisfied that devices of different makes will coordinate as required.

▼ **Figure 3.7.4** Typical location of a coordinated set of SPDs

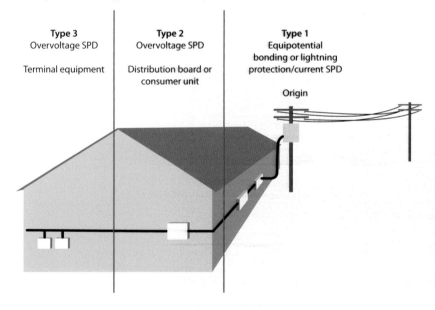

3.7.5 Critical length of connecting conductors for SPDs

534.2.9 To gain maximum protection the connecting conductors to SPDs must be kept as short as possible, to minimize additive inductive voltage drops across the conductors. The total lead length ($a + b$) should preferably not exceed 0.5 m but in no case exceed 1.0 m; see Figure 3.7.5.

Refer to the SPD manufacturer's instructions for optimal installation.

▼ **Figure 3.7.5** Critical length of connecting conductors for SPDs

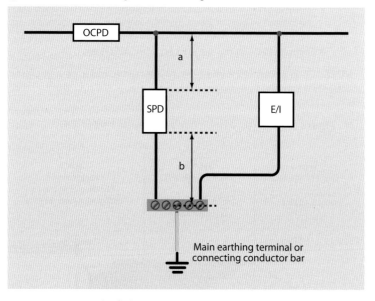

OCPD = overcurrent protective device
SPD = surge protective device
E/I = equipment or installation to be protected against overvoltages

3.7.6 Methods of connection

Primarily, the installation of SPDs must follow the manufacturer's instructions but minimum SPD connections at the origin of the electrical supply are usually made as those shown in Figure 3.7.6(i) (TN-C-S, TN-S, TT) and Figure 3.7.6(ii) (TT – SPDs upstream of RCD):

Type 1 SPDs should be installed upstream from any RCD to avoid unwanted tripping. Where this cannot be avoided, the RCD should be of the time-delayed or S-type.

▼ **Figure 3.7.6(i)** SPDs on load side of RCD

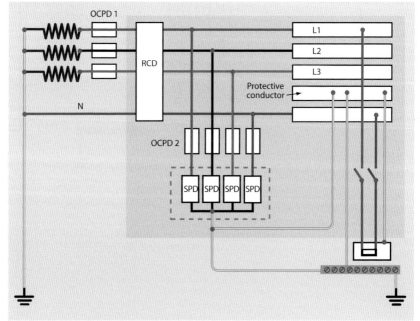

534.2.5(ii) ▼ **Figure 3.7.6(ii)** SPDs on supply side of RCD

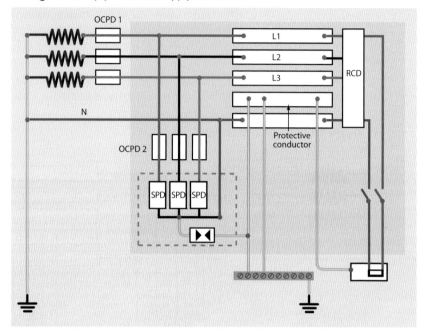

NOTE: See Appendix 16 of BS 7671:2008(2011) for further information regarding the connection of SPDs.

Earthing and bonding   4

4.1 Protective earthing

The purpose of protective earthing is to ensure that, in the event of a fault, such as between a line conductor and an exposed-conductive-part, sufficient current flows to operate the protective device, i.e. fuse to blow, circuit-breaker to operate or RCD to operate, in the required time.

411.4.2
411.5.1
Every *exposed-conductive-part* (a conductive part of equipment that can be touched and which is not a live part but which may become live under fault conditions) shall be connected by a protective conductor to the main earthing terminal and, hence, the means of earthing for the installation.

4.2 Legal requirements

ESQCR
2002
SI 2665
The Electricity Safety, Quality and Continuity Regulations 2002 (ESQCR), require that a distributor of electricity makes the supply neutral conductor or protective conductor available for the connection of the consumer's protective conductor where it can be reasonably concluded that such a connection is appropriate. Such a connection may be deemed inappropriate where there is a risk of the loss of the PEN conductor, for example, where bare overhead low voltage distribution cables supply a rural building. In such cases, an earth electrode must be provided and the installation will then form part of a TT system.

Essentially, permission to connect the consumer's protective conductor to the distributor's neutral can be denied to new installations but, where permission is granted, the distributor should maintain the connection.

NOTE: For some rural installations supplied by a PME arrangement, it may be pertinent to install an additional earth electrode to mitigate the effects of a PEN conductor becoming open-circuit; see IET Guidance Note 5.

4.3 Main protective bonding of metallic services

(Figures 2.1(i) to 2.1(iii))

The purpose of protective equipotential bonding is to reduce the voltages between the various exposed-conductive-parts and extraneous-conductive-parts of an installation, during a fault to earth and in the event of a fault on the distributor's network.

411.3.1.2 Part 2   Main protective bonding conductors are required to connect extraneous-conductive-parts to the main earthing terminal of the installation. An *extraneous-conductive-part* is a conductive part, such as a metal pipe, liable to introduce earth potential into the installation or building. It is common, particularly under certain fault conditions on the LV supply network, for a potential to exist between true earth, i.e. the general mass of Earth and the earth of the electrical system. Therefore, buried metallic parts which enter the building are to be bonded to the main earthing terminal of the electrical installation.

Examples of extraneous-conductive-parts are:

- i   metallic installation pipes
- ii   metallic gas installation pipes
- iii   other installation pipework, for example, heating oil
- iv   structural steelwork of the building where rising from the ground
- v   lightning protection systems (where required by BS EN 62305).

It is also necessary to consider not just metallic supply pipework but also internal metallic pipework which may have been buried in the ground for convenience, for example, central heating pipework cast into the concrete or buried in the screed of a floor at ground level. Such metallic pipes would normally be considered to be extraneous-conductive-parts.

4.4 Earthing conductor and main protective bonding conductor cross-sectional areas

The minimum cross-sectional areas (csa) of the earthing conductor and main protective bonding conductors are given in Table 4.4(i). For TT supplies, refer to Table 4.4(ii).

▼ **Table 4.4(i)**   Earthing conductor and main protective bonding conductor sizes (copper equivalent) for TN-S and TN-C-S supplies

	Line conductor or neutral conductor of PME supplies	mm^2	4	6	10	16	25	35	50	70
542.3 543.1	Earthing conductor not buried or buried and protected against corrosion and mechanical damage – see notes	mm^2	6	6	10	16	16	16	25	35
544.1.1	Main protective bonding conductor – see notes	mm^2	6	6	6	10	10	10	16	25
Table 54.8	Main protective bonding conductor for PME supplies (TN-C-S)	mm^2	10	10	10	10	10	10	16	25

Notes:

543.2.4   1   Protective conductors (including earthing and bonding conductors) of 10 mm² cross-sectional area or less shall be copper.

		2	The distributor may require a minimum size of earthing conductor at the origin of the supply of 16 mm² copper or greater for TN-S and TN-C-S supplies.

Table 54.7

542.3.1   **3**   Buried earthing conductors must be at least:
Table 54.1
- 25 mm² copper if not protected against corrosion
- 50 mm² steel if not protected against corrosion
- 16 mm² copper if not protected against mechanical damage but protected against corrosion
- 16 mm² coated steel if not protected against mechanical damage but protected against corrosion.

4   The distributor should be consulted when in doubt.

▼ **Table 4.4(ii)**   Copper earthing conductor cross-sectional area (csa) for TT supplies

Buried			Not buried		
Unprotected	Protected against corrosion	Protected against corrosion and mechanical damage	Unprotected	Protected against corrosion	Protected against corrosion and mechanical damage
mm²	mm²	mm²	mm²	mm²	mm²
25	16	2.5	4	4	2.5

Notes:

1 Assuming protected against corrosion by a sheath.

544.1.1   **2**   The main protective bonding conductors shall have a cross-sectional area of not less than half that required for the earthing conductor and not less than 6 mm².

Note that:

543.2.4   **i**   only copper conductors should be used; copper covered aluminium conductors or aluminium conductors or structural steel can only be used if special precautions outside the scope of this Guide are taken

544.1.2   **ii**   bonding connections to incoming metal services should be made as near as practicable to the point of entry of the services into the premises, but on the consumer's side of any insulating section

544.1.2   **iii**   where practicable, the connection to the gas, water, oil, etc., service should be within 600 mm of the service meter, or at the point of entry to the building if the service meter is external and must be on the consumer's side before any branch pipework and after any insulating section in the service. The connection must be made to hard pipework, not to soft or flexible meter connections

542.3.2   **iv**   the connection must be made using clamps (to BS 951) and be suitably protected against corrosion at the point of contact.

4.5   Main protective bonding of plastic services

There is no requirement to main bond an incoming service where the incoming service pipe is plastic, for example, where yellow is used for natural gas and blue for potable water.

Where there is a plastic incoming service and a metal installation within the premises, main bonding is recommended unless it has been confirmed that any metallic pipework within the building is not introducing earth potential (see 4.3).

544.1.2 All main bonding connections are to be applied to the consumer's side of any meter, main stop valve or insulating insert and, where practicable, within 600 mm of the meter outlet union or point of entry to the building if the meter is external.

4.6 Supplementary equipotential bonding

The purpose of supplementary equipotential bonding is to reduce the voltage between the various exposed-conductive-parts and extraneous-conductive-parts of a location during a fault to earth.

NOTE: Where a required disconnection time cannot be achieved, supplementary bonding must be applied, however, this is outside the scope of this Guide. See Regulations 411.3.2.5 and 411.3.2.6 and Guidance Note 1.

The cross-sectional area of supplementary bonding conductors is given in Table 4.6.

▼ **Table 4.6** Supplementary bonding conductors

544.2

Size of circuit protective conductor (mm²)	Minimum cross-sectional area of supplementary bonding conductor (mm²)					
	Exposed-conductive-part to extraneous-conductive-part		Exposed-conductive-part to exposed-conductive-part		Extraneous-conductive-part to extraneous-conductive-part*	
	mechanically protected	not mechanically protected	mechanically protected	not mechanically protected	mechanically protected	not mechanically protected
	1	2	3	4	5	6
1.0	1.0	4.0	1.0	4.0	2.5	4.0
1.5	1.0	4.0	1.5	4.0	2.5	4.0
2.5	1.5	4.0	2.5	4.0	2.5	4.0
4.0	2.5	4.0	4.0	4.0	2.5	4.0
6.0	4.0	4.0	6.0	6.0	2.5	4.0
10.0	6.0	6.0	10.0	10.0	2.5	4.0
16.0	10.0	10.0	16.0	16.0	2.5	4.0

544.2.3 * If one of the extraneous-conductive-parts is connected to an exposed-conductive-part, the bonding conductor must be no smaller than that required by column 1 or 2.

4.7 Additional protection – supplementary equipotential bonding

415.2 Supplementary equipotential bonding is required in some of the locations and installations falling within the scope of Part 7 of BS 7671.

If the installation meets the requirements of BS 7671:2008(2011) for earthing and bonding, there is no specific requirement for supplementary equipotential bonding of:

- kitchen pipes, sinks or draining boards
- metallic boiler pipework
- metallic furniture in kitchens
- metallic pipes to wash-hand basins and WCs
- locations containing a bath or shower, providing the conditions of Regulation 701.415.2 are met.

NOTE: Metallic waste pipes deemed to be extraneous-conductive-parts must be connected by main protective bonding conductors to the main earthing terminal; see also 4.3.

4.8 Supplementary bonding of plastic pipe installations

Supplementary bonding is not required to metallic parts supplied by plastic pipes, for example, radiators, kitchen sinks or bathroom taps.

4.9 Earth electrode

This is connected to the main earthing terminal by the earthing conductor and provides part of the earth fault loop path for a TT installation; see Figure 2.1(iii).

It is recommended that the earth fault loop impedance for TT installations does not exceed 200 Ω.

Metallic gas or water utility or other metallic service pipes are not to be used as an earth electrode, although they must be bonded if they are extraneous-conductive-parts; see also 4.3.

NOTE: Regulation 542.2.6 permits the use of privately owned water supply pipework for use as an earth electrode where precautions are taken against its removal and it has been considered for such use. This relaxation will not apply to an installation within a dwelling.

4.10 Types of earth electrode

The following types of earth electrode are recognised:

i earth rods or pipes
ii earth tapes or wires
iii earth plates
iv underground structural metalwork embedded in foundations
v welded metal reinforcement of concrete embedded in the ground (excluding pre-stressed concrete)

542.2.5	**vi** lead sheaths and metal coverings of cables, which must meet all the following conditions:
	    a adequate precautions to prevent excessive deterioration by corrosion
	    b the sheath or covering shall be in effective contact with Earth
	    c the consent of the owner of the cable shall be obtained
	    d arrangements shall exist for the owner of the electrical installation to be warned of any proposed change to the cable which might affect its suitability as an earth electrode.

4.11 Typical earthing arrangements for various types of earthing system

Figures 2.1(i) to 2.1(iii) show single-phase arrangements but three-phase arrangements are similar.

Table 54.7 Table 54.8 544.1.1	The protective conductor sizes as shown in Figures 2.1(i) to 2.1(iii) refer to copper conductors and are related to 25 mm² supply tails from the meter.
542.3.1 543.1.3	For TT systems protected by an RCD with an earth electrode resistance 1 ohm or greater, the earthing conductor size need not exceed 2.5 mm² if protected against corrosion by a sheath and if also protected against mechanical damage; otherwise, see Table 4.4(ii).
542.4.2	The earthing bar is sometimes used as the main earthing terminal, however, means must be provided in an accessible position for disconnecting the earthing conductor to facilitate measurement of external earth fault loop impedance, Ze.

NOTE: For TN-S and TN-C-S installations, advice about the availability of an earthing facility and the precise arrangements for connection should be obtained from the distributor or supplier.

Isolation and switching 5

5.1 Isolation
537.2

5.1.1 Requirement
132.15.1

Means of isolation should be provided:

i at the origin of the installation

537.1.4 A main linked switch or circuit-breaker should be provided as a means of isolation and of interrupting the supply on load.

For single-phase household and similar supplies that may be operated by unskilled persons, a double-pole device must be used for both TT and TN systems.

For a three-phase supply to an installation forming part of a TT system, an isolator must interrupt the line and neutral conductors. In a TN-S or TN-C-S system only the line conductors need be interrupted.

ii for every circuit

537.2.1.1 Other than at the origin of the installation, every circuit or group of circuits that may have to be isolated without interrupting the supply to other circuits should be provided with its own isolating device. The device must switch all live conductors in a TT system and all line conductors in a TN system.

537.2.1.2 **iii for every item of equipment**

iv for every motor

132.15.2 Every fixed electric motor should be provided with a readily accessible and easily operated device to switch off the motor and all associated equipment including any automatic circuit-breaker. The device must be so placed as to prevent danger.

537.1.3 **v for every supply.**

5.1.2 The switchgear

537.2.2.2 The position of the contacts of the isolator must either be externally visible or be clearly, positively and reliably indicated.

537.2.2.3 The device must be designed or installed to prevent unintentional or inadvertent closure.

Each device used for isolation must be clearly identified by position or durable marking to indicate the installation or circuit that it isolates.

On-Site Guide | **45**
© The Institution of Engineering and Technology

537.2.1.5	If it is installed remotely from the equipment to be isolated, the device must be capable of being secured in the OPEN position.

Guidance on the selection of devices for isolation is given in Appendix J.

537.3 5.2 Switching off for mechanical maintenance

537.3.1.1	A means of switching off for mechanical maintenance is required where mechanical maintenance may involve a risk of injury – for example, from mechanical movement of machinery or hot items when replacing lamps.
537.3.1.2	The means of switching off for mechanical maintenance must be able to be made secure to prevent electrically powered equipment from becoming unintentionally started during the mechanical maintenance, unless the means of switching off is continuously under the control of the person performing the maintenance.

Each device for switching off for mechanical maintenance must:

537.3.2.1	i	where practicable, be inserted in the main supply circuit
537.3.2.1	ii	be capable of switching the full load current
537.3.2.2	iii	be manually operated
537.3.2.2	iv	have either an externally visible contact gap or a clearly and reliably indicated OFF position. An indicator light should not be relied upon
537.3.2.3	v	be designed and/or installed so as to prevent inadvertent or unintentional switching on
537.3.2.4	vi	be installed and durably marked so as to be readily identifiable and convenient for use.

537.3.2.6	A plug and socket-outlet or similar device of rating not exceeding 16 A may be used for switching off for mechanical maintenance.

537.4 5.3 Emergency switching

537.4.1.1 537.4.1.2	An emergency switch is to be provided for any part of an installation where it may be necessary to control the supply in order to remove an unexpected danger. Where there is a risk of electric shock the emergency switch is to disconnect all live conductors, except in three-phase TN-S and TN-C-S systems where the neutral need not be switched.
537.4.1.3	The means of emergency switching must act as directly as possible on the appropriate supply conductors and the arrangement must be such that one single action only will interrupt the appropriate supply.
537.4.2.8	A plug and socket-outlet or similar device must not be selected as a device for emergency switching.

An emergency switch must be:

537.4.2.1	i	capable of cutting off the full load current, taking account of stalled motor currents where appropriate

537.4.2.3	ii	hand operated and directly interrupt the main circuit where practicable
537.4.2.4	iii	clearly identified, preferably by colour. If a colour is used, this should be red with a contrasting background
537.4.2.5	iv	readily accessible at the place where danger might occur and, where appropriate, at any additional remote position from which that danger can be removed
537.4.2.6	v	of the latching type or capable of being restrained in the 'OFF' or 'STOP' position, unless both the means of operation and re-energizing are under the control of the same person. The release of an emergency switching device must not re-energize the relevant part of the installation; it must be necessary to take a further action, such as pushing a 'start' button
537.4.2.7	vi	so placed and durably marked so as to be readily identifiable and convenient for its intended use.

5.4 Functional switching

537.5

537.5.1.1 A switch must be installed in each part of a circuit which may require to be controlled independently of other parts of the installation.

530.3.2 Switches must not be installed in the neutral conductor alone.

537.5.1.3 All current-using equipment requiring control shall be controlled by a switch.

537.5.2.3 Off-load isolators, fuses and links must not be used for functional switching.

NOTE: Table 53.4 of BS 7671:2008(2011) permits the use of circuit-breakers for functional switching purposes but, in each case, the manufacturer should be consulted to establish suitability.

5.5 Firefighter's switch

537.6

537.6.1 A firefighter's switch must be provided to disconnect the supply to any exterior electrical installation operating at a voltage exceeding low voltage, for example, a neon sign or any interior discharge lighting installation operating at a voltage exceeding low voltage.

NOTE: Such installations are outside the scope of this Guide; see Regulations 537.6.1 to 537.6.4 of BS 7671:2008(2011).

Labelling 6

The following durable labels are to be securely fixed on or adjacent to installed equipment.

6.1 Additional protection

411.3.3 (b)   A specific labelled or otherwise suitably identified socket-outlet provided for connection of a particular item of equipment.

6.2 Retention of a dangerous electrical charge

416.2.5   If, behind a barrier or within an enclosure, an item of equipment such as a capacitor is installed which may retain a dangerous electrical charge after it has been switched off, a warning label must be provided. Small capacitors such as those used for arc extinction and for delaying the response of relays, etc., are not considered dangerous.

NOTE: Unintentional contact is not considered dangerous if the voltage resulting from static charge falls below 120 V d.c. in less than 5 s after disconnection from the power supply.

6.3 Where the operator cannot observe the operation of switchgear and controlgear

514.1.1   Except where there is no possibility of confusion, a label or other suitable means of identification must be provided to indicate the purpose of each item of switchgear and controlgear. Where the operator cannot observe the operation of switchgear and controlgear and where this might lead to danger, a suitable indicator complying, where applicable, with BS EN 60073 and BS EN 60447, should be fixed in a position visible to the operator.

6.4 Unexpected presence of nominal voltage exceeding 230 V

514.10.1   Where a nominal voltage exceeding 230 V to earth exists and it would not normally be expected, a warning label stating the maximum voltage present must be provided where it can be seen before gaining access to live parts.

6

6.5 Connection of earthing and bonding conductors

514.13.1 A permanent label to BS 951 (Figure 6.5) must be permanently fixed in a visible position at or near the point of connection of:

    i   every earthing conductor to an earth electrode,
    ii   every protective bonding conductor to extraneous-conductive-parts, and
    iii at the main earth terminal, where it is not part of the main switchgear.

▼ **Figure 6.5**   Label at connection of earthing and bonding conductors

6.6 Purpose of switchgear and controlgear

514.1.1 Unless there is no possibility of confusion, a label indicating the purpose of each item of switchgear and controlgear must be fixed on or adjacent to the gear. It may be necessary to label the item controlled, in addition to its controlgear.

6.7 Identification of protective devices

514.8.1 A protective device, for example, a fuse or circuit-breaker, must be arranged and identified so that the circuit protected may be easily recognised.

6.8 Identification of isolators

537.2.2.6 Where it is not immediately apparent, all isolating devices must be clearly identified by position or durable marking. The location of each disconnector or isolator must be indicated unless there is no possibility of confusion.

6.9 Isolation requiring more than one device

514.11.1 A durable warning notice must be permanently fixed in a clearly visible position to identify the appropriate isolating devices, where equipment or an enclosure contains live parts which cannot be isolated by a single device.

6.10 Periodic inspection and testing

514.12.1 A notice of durable material indelibly marked with the words as Figure 6.10 must be fixed in a prominent position at or near the origin of every installation. The person carrying out the initial verification must complete the notice and it must be updated after each periodic inspection.

▼ **Figure 6.10** Label for periodic inspection and testing

> **IMPORTANT**
>
> This installation should be periodically inspected and tested and a report on its condition obtained, as prescribed in the IET Wiring Regulations BS 7671 Requirements for Electrical Installations.
>
> Date of last inspection ....
>
> Recommended date of next inspection ....

6.11 Diagrams

514.9.1 A diagram, chart or schedule must be provided indicating:

   i   the number of points, size and type of cables for each circuit,
   ii the method of providing protection against electric shock,
   iii information to identify devices for protection, isolation and switching, and
   iv any circuit or equipment vulnerable during a typical test, e.g. SELV power supply units of lighting circuits which could be damaged by an insulation test.

For simple installations, the foregoing information may be given in a schedule, with a durable copy provided within or adjacent to the distribution board or consumer unit.

6.12 Residual current devices

514.12.2 Where an installation incorporates an RCD, a notice with the words in Figure 6.12 (and no smaller than the example shown in BS 7671:2008(2011)) must be fixed in a permanent position at or near the origin of the installation.

▼ **Figure 6.12** Label for the testing of a residual current device

> This installation, or part of it, is protected by a device which automatically switches off the power supply if an earth fault develops. **Test quarterly** by pressing the button marked **'T'** or **'Test'**. The device should switch off the supply and should be then switched on to restore the supply. If the device does not switch off the supply when the button is pressed seek expert advice.

6.13 Warning notice – non-standard colours

514.14.1 If additions or alterations are made to an installation so that some of the wiring complies with the harmonized colours of Table K1 in Appendix K and there is also wiring in the earlier colours, a warning notice must be affixed at or near the appropriate distribution board with the wording in Figure 6.13.

▼ **Figure 6.13** Label advising of wiring colours to two versions of BS 7671

> **CAUTION**
> This installation has wiring colours to two versions of BS 7671.
> Great care should be taken before undertaking extension, alteration or repair that all conductors are correctly identified.

6.14 Warning notice – alternative supplies

514.15.1 Where an installation includes additional or alternative supplies, such as a PV installation, which is used as an additional source of supply in parallel with another source, normally the distributor's supply, warning notices must be affixed at the following locations in the installation:

    a at the origin of the installation
    b at the meter position, if remote from the origin

52 | On-Site Guide
© The Institution of Engineering and Technology

c at the consumer unit or distribution board to which the additional or alternative supply is connected
d at all points of isolation of all sources of supply.

The warning notice must have the wording in Figure 6.14.

▼ **Figure 6.14** Label advising of multiple supplies

6.15 Warning notice – high protective conductor current

543.7.1.105 At the distribution board, information must be provided indicating those circuits having a high protective conductor current. This information must be positioned so as to be visible to a person who is modifying or extending the circuit (Figure 6.15).

▼ **Figure 6.15** Label advising of high protective conductor current

WARNING
HIGH PROTECTIVE CONDUCTOR CURRENT
The following circuits have a high protective conductor current:

....

....

6.16 Warning notice – photovoltaic systems

712.537.2.2 All junction boxes (PV generator and PV array boxes) must carry a warning label indicating that parts inside the boxes may still be live after isolation from the PV convertor (Figure 6.16).

▼ **Figure 6.16** Label advising of live parts within enclosures in a PV system

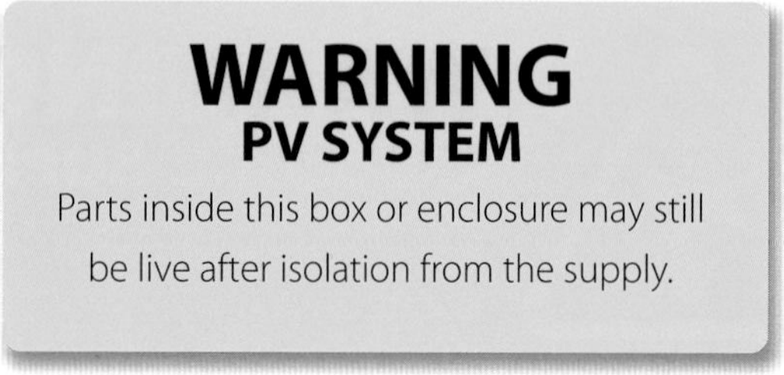

WARNING
PV SYSTEM
Parts inside this box or enclosure may still be live after isolation from the supply.

Final circuits 7

7.1 Final circuits

Table 7.1(i) has been designed to enable a radial or ring final circuit to be installed without calculation where the supply is at 230 V single-phase or 400 V three-phase. For other voltages, the maximum circuit length given in the table must be corrected by the application of the formula:

$$L_p = \frac{L_t \times U_0}{230}$$

where:

L_p is the permitted length for voltage U_0
L_t is the tabulated length for 230 V
U_0 is the supply voltage.

The conditions assumed are that:

i the installation is supplied by
 a a TN-C-S system with a typical external earth fault loop impedance, Z_e, of 0.35 Ω, or
 b a TN-S system with a typical Z_e of 0.8 Ω, or
 c a TT system with RCDs installed as described in 3.6
ii the final circuit is connected to a distribution board or consumer unit at the origin of the installation
iii the installation method is listed in column 4 of Table 7.1(i)
iv the ambient temperature throughout the length of the circuit does not exceed 30 °C
v the characteristics of protective devices are in accordance with Appendix 3 of BS 7671
vi the cable conductors are of copper
vii for other than lighting circuits, the voltage drop must not exceed 5 per cent
viii the following disconnection times are applicable:
 ▶ 0.4 s for circuits up to and including 32 A
 ▶ 5 s for circuits greater than 32 A.

7

▼ **Table 7.1(i)** Maximum cable length for a 230 V final circuit in domestic premises and similar using 70 °C thermoplastic (PVC) insulated and sheathed flat cable

Rating (A)	Protective device		Cable size (mm²)	Allowed installation methods (note 2)	Maximum length (m) (note 1)			
					$Z_e \leq 0.8\,\Omega$ TN-S		$Z_e \leq 0.35\,\Omega$ TN-C-S	
	Type				RCD 30 mA	No RCD	RCD 30 mA	No RCD
1	2		3	4	5	6	7	8
Ring final circuits (5% voltage drop, load distributed)								
30	BS 1361		2.5/1.5	100,102, A, C	111	59zs	111	111
30	BS 3036		2.5/1.5	100,102, A, C	111	49zs	111	111
30	BS 1361		4.0/1.5	100,101,102, A, C	183	69zs	183	159zs
30	BS 3036		4.0/1.5		183	57zs	183	147zs
32	BS 88-2.2, BS 88-6		2.5/1.5	100,102, A, C	106	41zs	106	106
32	cb/RCBO Type B				106	106	106	106
32	cb/RCBO Type C				NPsc	NPzs	82sc	63zs
32	cb/RCBO Type D				NPsc	NPzs	2sc	1zs
32	BS 88-2		2.5/1.5	100,102, A, C	106	41zs	106	106
32	BS 88-2		4.0/1.5	100,101,102, A, C	176	48zs	176	138zs
32	BS 88-3		2.5/1.5	100,102, A, C	106	27zs	106	104zs
32	BS 88-3		4.0/1.5	100,101,102, A, C	176	32zs	176	122zs
32	BS 88-2.2, BS 88-6		4.0/1.5	100,101,102, A, C	176	47zs	176	137zs
32	cb/RCBO Type B				176	127zs	176	176
32	cb/RCBO Type C				NFsc	NPzs	133sc	73zs
32	cb/RCBO Type D				NFsc	NPzs	3sc	1zs
Lighting circuits (3% voltage drop, load distributed)								
5	BS 1361		1.0/1.0	100,101,102,103, A, C	71	71	71	71
5	BS 3036				71	71	71	71

Table 7.1(i) continued

Lighting circuits (3% voltage drop, load distributed)

Rating (A)	Protective device Type	Cable size (mm²)	Allowed installation methods (note 2)	Maximum length (m) (note 1)			
				$Z_e \leq 0.8\,\Omega$ TN-S		$Z_e \leq 0.35\,\Omega$ TN-C-S	
				RCD 30 mA	No RCD	RCD 30 mA	No RCD
1	2	3	4	5	6	7	8
5	BS 1361 BS 3036	1.5/1.0	100,101,102,103, A, C	108 108	108 108	108 108	108 108
5	BS 88-3	1.0/1.0	100,101,102,103, A, C	71	71	71	71
5	BS 88-3	1.5/1.0	100,101,102,103, A, C	108	108	108	108
6	BS 88-2.2, BS 88-6 cb/RCBO Type B cb/RCBO Type C cb/RCBO Type D	1.0/1.0	100,101,102,103, A, C	59 59 59 25sc	59 59 59 25zs	59 59 59 36sc	59 59 59 36zs
6	BS 88-2	1.0/1.0	100,101,102,103, A, C	59	59	59	59
6	BS 88-2.2, BS 88-6 cb/RCBO Type B cb/RCBO Type C cb/RCBO Type D	1.5/1.0	100,101,102,103, A, C	90 90 90 38sc	90 90 83zs 30zs	90 90 90 53sc	90 90 90 43zs
6	BS 88-2	1.5/1.0	100,101,102,103, A, C	90	90	90	90
10	BS 88-2.2, BS 88-6 cb/RCBO Type B cb/RCBO Type C cb/RCBO Type D	1.0/1.0	100, 101, 102, A, C	35 35 34sc 8sc	35 35 34zs 8zs	35 35 35 18sc	35 35 35 18zs
10	BS 88-2.2, BS 88-6 cb/RCBO Type B cb/RCBO Type C cb/RCBO Type D	1.5/1.0	100, 101, 102, A, C	52 52 51sc 12sc	52 52 41zs 9zs	52 52 52 27sc	52 52 52 22zs

▼ Table 7.1(i) continued

Protective device		Cable size (mm²)	Allowed installation methods (note 2)	Maximum length (m) (note 1)			
				$Z_e \leq 0.8\,\Omega$ TN-S		$Z_e \leq 0.35\,\Omega$ TN-C-S	
Rating (A)	Type			RCD 30 mA	No RCD	RCD 30 mA	No RCD
1	2	3	4	5	6	7	8
Lighting circuits (3% voltage drop, load distributed)							
10	BS 88-2	1.0/1.0	100, 101, 102, A, C	35	35	35	35
10	BS 88-2	1.5/1.0	100, 101, 102, A, C	52	52	52	52
15	BS 1361	1.5/1.0	100, 102, C	36	36	36	36
15	BS 1361	2.5/1.5	100, 101, 102, A, C	58	58	58	58
16	BS 88-2.2, BS 88-6	1.5/1.0	100, 102, C	33	33	33	33
	cb/RCBO Type B			33	33	33	33
	cb/RCBO Type C			21sc	17zs	33	30zs
	cb/RCBO Type D			NPsc	NPzs	12sc	10zs
16	BS 88-2.2, BS 88-6	2.5/1.5	100, 101, 102, A, C	53	53	53	53
	cb/RCBO Type B			53	53	53	53
	cb/RCBO Type C			35sc	27zs	53	46zs
	cb/RCBO Type D			NPsc	NPzs	20sc	15zs
16	BS 88-2	1.5/1.0	100, 102, C	33	33	33	33
16	BS 88-2	2.5/1.5	100, 101, 102, A, C	53	53	53	53
16	BS 88-3	1.5/1.0	100, 102, C	33	33	33	33
16	BS 88-3	2.5/1.5	100, 101, 102, A, C	53	53	53	53
Radial final circuits (5% voltage drop, terminal load)							
5	BS 1361 / BS 3036	1.0/1.0	100, 101, 102, 103, A, C	56	56	56	56
				56	56	56	56
5	BS 88-3	1.0/1.0	100, 101, 102, 103, A, C	56	56	56	56

▼ Table 7.1(i) continued

Protective device		Cable size (mm²)	Allowed installation methods (note 2)	Maximum length (m) (note 1)			
				$Z_e \leq 0.8\,\Omega$ TN-S		$Z_e \leq 0.35\,\Omega$ TN-C-S	
Rating (A)	Type			RCD 30 mA	No RCD	RCD 30 mA	No RCD
1	2	3	4	5	6	7	8
Radial final circuits (5% voltage drop, terminal load)							
5	BS 1361 BS 3036	1.5/1.0	100, 101, 102, 103, A, C	88 88	88 88	88 88	88 88
5	BS 88-3	1.5/1.0	100, 101, 102, 103, A, C	88	88	88	88
6	BS 88-2.2, BS 88-6 cb/RCBO Type B cb/RCBO Type C cb/RCBO Type D	1.0/1.0	100, 101, 102, 103, A, C	46 46 46 25sc	46 46 46 25zs	46 46 46 36sc	46 46 46 36zs
6	BS 88-2	1.0/1.0	100, 101, 102, 103, A, C	46	46	46	46
6	BS 88-2	1.5/1.0	100, 101, 102, 103, A, C	72	72	72	72
6	BS 88-2.2, BS 88-6 cb/RCBO Type B cb/RCBO Type C cb/RCBO Type D	1.5/1.0	100, 101, 102, 103, A, C	72 72 72 38sc	72 72 72 30zs	72 72 72 53sc	72 72 72 43zs
10	BS 88-2.2, BS 88-6 cb/RCBO Type B cb/RCBO Type C cb/RCBO Type D	1.0/1.0	100, 101, 102, A, C	26 26 26 8sc	26 26 26 8zs	26 26 26 18sc	26 26 26 18zs
10	BS 88-2.2, BS 88-6 cb/RCBO Type B cb/RCBO Type C cb/RCBO Type D	1.5/1.0	100, 101, 102, 103, A, C	39 39 39 12sc	39 39 39 9zs	39 39 39 27sc	39 39 39 22zs
10	BS 88-2	1.0/1.0	100, 101, 102, A, C	26	26	26	26

▼ Table 7.1(i) *continued*

Rating (A)	Protective device		Cable size (mm²)	Allowed installation methods (note 2)	Maximum length (m) (note 1)			
					$Z_e \leq 0.8\,\Omega$ TN-S		$Z_e \leq 0.35\,\Omega$ TN-C-S	
					RCD 30 mA	No RCD	RCD 30 mA	No RCD
	Type							
1	2		3	4	5	6	7	8
Radial final circuits (5% voltage drop, terminal load)								
10	BS 88-2		1.5/1.0	100, 101, 102, 103, A, C	39	39	39	39
15	BS 1361		1.0/1.0	C	17	17	17	17
	BS 3036			NP	NPol	NPol	NPol	NPol
15	BS 1361		1.5/1.0	100,102,C	26	26	26	26
	BS 3036			NP	NPol	NPol	NPol	NPol
15	BS 1361		2.5/1.5	100, 101, 102, A, C	43	43	43	43
	BS 3036			100,102, C	45	45	45	45
15	BS 1361		4.0/1.5	100, 101, 102, 103, A, C	72	72	72	72
	BS 3036			100, 101,102, A, C	75	75	75	75
16	BS 88-2.2, BS 88-6		1.0/1.0	C	16	16	16	16
	cb/RCBO Type B				16	16	16	16
	cb/RCBO Type C				14sc	14zs	16	16
	cb/RCBO Type D				NPsc	NPzs	8sc	8zs
16	BS 88-2		1.5/1.0	100, 102, C	24	24	24	24
16	BS 88-2		2.5/1.5	100, 101, 102, A, C	40	40	40	40
16	BS 88-2		4.0/1.5	100, 101, 102, 103, A, C	66	66	66	66
16	BS 88-3		1.5/1.0	100, 102, C	24	24	24	24
16	BS 88-3		2.5/1.5	100, 101, 102, A, C	40	40	40	40
16	BS 88-3		4.0/1.5	100, 101, 102, 103, A, C	66	66	66	66

▼ **Table 7.1(i)** *continued*

Rating (A)	Protective device		Cable size (mm²)	Allowed installation methods (note 2)	Maximum length (m) (note 1)			
					$Z_e \leq 0.8\ \Omega$ TN-S		$Z_e \leq 0.35\ \Omega$ TN-C-S	
		Type			RCD 30 mA	No RCD	RCD 30 mA	No RCD
1	2		3	4	5	6	7	8
Radial final circuits (5% voltage drop, terminal load)								
16	BS 88-2.2, BS 88-6		1.5/1.0	100, 102, C	24	24	24	24
	cb/RCBO Type B				24	24	24	24
	cb/RCBO Type C				21sc	17zs	24	24
	cb/RCBO Type D				NPsc	NPzs	12sc	10zs
16	BS 88-2.2, BS 88-6		2.5/1.5	100, 101, 102, A, C	40	40	40	40
	cb/RCBO Type B				40	40	40	40
	cb/RCBO Type C				35sc	27zs	40	40
	cb/RCBO Type D				NPsc	NPzs	20sc	15zs
16	BS 88-2.2, BS 88-6		4.0/1.5	100, 101, 102, 103, A, C	66	66	66	66
	cb/RCBO Type B				66	66	66	66
	cb/RCBO Type C				57sc	31zs	66	54zs
	cb/RCBO Type D				NPsc	NPzs	33sc	18zs
20	BS 88-2.2, BS 88-6		2.5/1.5	100, 102, A, C	31	31	31	31
	BS 1361			100, 102, A, C	31	31	31	31
	BS 3036			NP	NPol	NPol	NPol	NPol
	cb/RCBO Type B			100, 102, A, C	31	31	31	31
	cb/RCBO Type C				19sc	14zs	31	31
	cb/RCBO Type D				NPsc	NPzs	12sc	9zs
20	BS 88-2.2, BS 88-6		4.0/1.5	100, 101, 102, A, C	53	48zs	53	53
	BS 1361			100, 101, 102, A, C	53	44zs	53	53
	BS 3036			C	57	48zs	57	57
	cb/RCBO Type B				53	53	53	53
	cb/RCBO Type C			100, 101, 102, A, C	31sc	17zs	53	39zs
	cb/RCBO Type D				NPsc	NPzs	20sc	11zs

▼ Table 7.1(i) *continued*

Rating (A)	Protective device Type	Cable size (mm²)	Allowed installation methods (note 2)	Maximum length (m) (note 1)			
				$Z_e \leq 0.8\,\Omega$ TN-S		$Z_e \leq 0.35\,\Omega$ TN-C-S	
				RCD 30 mA	No RCD	RCD 30 mA	No RCD
1	2	3	4	5	6	7	8
Radial final circuits (5% voltage drop, terminal load)							
20	BS 88-2.2, BS 88-6	6.0/2.5	100, 101,102, 103, A, C	81	77zs	81	81
	BS 1361		100, 101,102, 103, A, C	81	71zs	81	81
	BS 3036		100, 102, A, C	85	77zs	85	85
	cb/RCBO Type B			81	81	81	81
	cb/RCBO Type C		} 100, 101,102, 103, A, C	47sc	27zs	81	63zs
	cb/RCBO Type D			NPsc	NPzs	30sc	17zs
20	BS 88-2	2.5/1.5	100, 102, A, C	31	31	31	31
20	BS 88-2	4.0/1.5	100, 101, 102, A, C	48sc	48sc	53	53
20	BS 88-2	6.0/2.5	100, 101,102, 103, A, C	77sc	77sc	81	81
20	BS 88-3	2.5/1.5	100, 102, A, C	31	31	31	31
20	BS 88-3	4.0/1.5	100, 101, 102, A, C	53	53	53	53
20	BS 88-3	6.0/2.5	100, 101,102, 103, A, C	81	81	81	81
25	BS 88-2.2, BS 88-6	2.5/1.5	} C	26	26	26	26
	cb/RCBO Type B			26	26	26	26
	cb/RCBO Type C			6sc	5zs	26	24zs
	cb/RCBO Type D			NPsc	NPzs	6sc	4zs
25	BS 88-2.2, BS 88-6	4.0/1.5	} 100, 102, A, C	42	31zs	42	42
	cb/RCBO Type B			42	42	42	42
	cb/RCBO Type C			10sc	6zs	42	28zs
	cb/RCBO Type D			NPsc	NPzs	9sc	5zs

▼ Table 7.1(i) continued

Rating (A)	Protective device Type	Cable size (mm²)	Allowed installation methods (note 2)	Maximum length (m) (note 1)			
				$Z_e \leq 0.8\,\Omega$ TN-S	No RCD	$Z_e \leq 0.35\,\Omega$ TN-C-S	No RCD
				RCD 30 mA		RCD 30 mA	
1	2	3	4	5	6	7	8
Radial final circuits (5% voltage drop, terminal load)							
25	BS 88-2.2, BS 88-6	6.0/2.5	100, 101, 102, A, C	64	50zs	64	64
	cb/RCBO Type B			64	64	64	64
	cb/RCBO Type C			16sc	9zs	64	45zs
	cb/RCBO Type D			NPsc	NPzs	14sc	8zs
25	BS 88-2	2.5/1.5	C	26	23zs	26	26
25	BS 88-2	4.0/1.5	100, 102, A, C	42	27zs	42	42
25	BS 88-2	6.0/2.5	100, 101, 102, A, C	64	44zs	64	64
30	BS 1361	4.0/1.5	C	36	17zs	36	36
	BS 3036		NP	NPol	NPol	NPol	NPol
30	BS 1361	6.0/2.5	100, 102, A, C	53	27zs	53	53
	BS 3036		C	57	23zs	57	57
30	BS 1361	10.0/4.0	100, 101, 102, 103, A, C	90	45zs	90	90
	BS 3036		100, 102, A, C	93	37zs	93	93
32	BS 88-2.2, BS 88-6	4.0/1.5	C	33	11zs	33	33
	cb/RCBO Type B			33	31zs	33	33
	cb/RCBO Type C			NPsc	NPzs	33	18zs
	cb/RCBO Type D			NPsc	NPzs	1sc	NPzs
32	BS 88-2.2, BS 88-6	6.0/2.5	100,102, A, C	49	19zs	49	49
	cb/RCBO Type B			49	49	49	49
	cb/RCBO Type C			NPsc	NPzs	49	29zs
	cb/RCBO Type D			NPsc	NPzs	1sc	1zs

▼ Table 7.1(i) *continued*

Rating (A)	Protective device		Cable size (mm²)	Allowed installation methods (note 2)	Maximum length (m) (note 1)			
	Type				$Z_e \leq 0.8\,\Omega$ TN-S		$Z_e \leq 0.35\,\Omega$ TN-C-S	
					RCD 30 mA	No RCD	RCD 30 mA	No RCD
1	2		3	4	5	6	7	8
Radial final circuits (5% voltage drop, terminal load)								
32	BS 88-2.2, BS 88-6		10.0/4.0	100, 101, 102, 103, A, C	81	31zs	81	81
	cb/RCBO Type B				81	81	81	81
	cb/RCBO Type C				NPsc	NPzs	81	47zs
	cb/RCBO Type D				NPsc	NPzs	2sc	1zs
32	BS 88-2		4.0/1.5	C	33	12zs	33	33
32	BS 88-2		6.0/2.5	100, 102, A, C	49	19zs	49	49
32	BS 88-2		10/4.0	100, 101, 102, 103, A, C	81	31zs	81	81
32	BS 88-3		4.0/1.5	C	33	8zs	33	30zs
32	BS 88-3		6.0/2.5	100, 102, A, C	49	13zs	49	48zs
32	BS 88-3		10/4.0	100, 101, 102, 103, A, C	81	21zs	81	79zs
40	BS 88-2.2, BS 88-6		6.0/2.5	C	43	43	43	43
	cb/RCBO Type B				43	20zs	43	43
	cb/RCBO Type C				NPsc	NPzs	30sc	17zs
	cb/RCBO Type D				NPsc	NPzs	NPsc	NPzs
40	BS 88-2.2, BS 88-6		10.0/4.0	100, 102, A, C	66	66	66	66
	cb/RCBO Type B				66	45zs	66	66
	cb/RCBO Type C				NPsc	NPzs	51sc	29zs
	cb/RCBO Type D				NPsc	NPzs	NPsc	NPzs
40	BS 88-2.2, BS 88-6		16.0/6.0	100, 101, 102, 103, A, C	104	104	104	104
	cb/RCBO Type B				104	68zs	104	104
	cb/RCBO Type C				NPsc	NPzs	81sc	44zs
	cb/RCBO Type D				NPsc	NPzs	NPsc	NPzs

▼ Table 7.1(i) continued

Rating (A)	Protective device Type	Cable size (mm²)	Allowed installation methods (note 2)	Maximum length (m) (note 1)			
				$Z_e \leq 0.8\ \Omega$ TN-S		$Z_e \leq 0.35\ \Omega$ TN-C-S	
				RCD 30 mA	No RCD	RCD 30 mA	No RCD
1	2	3	4	5	6	7	8
Radial final circuits (5% voltage drop, terminal load)							
40	BS 88-2	6.0/2.5	C	43	NPzs	43	35zs
40	BS 88-2	10.0/4.0	100, 102, A, C	66	NPzs	66	57zs
40	BS 88-2	16.0/6.0	100, 101, 102, 103, A, C	104	NPzs	104	87zs
45	BS 1361 BS 3036	6.0/2.5	C NP	21sc NPol	NPad NPol	35 NPol	22ad NPol
45	BS 1361 BS 3036	10.0/4.0	100, 102, C C	36sc 62	5ad 62	58 62	58 62
45	BS 1361 BS 3036	16.0/6.0	100, 101, 102, A, C 102, C	58sc 97	31zs 97	91 97	91 97
45	BS 88-3	6.0/2.5	C	35	NPzs	35	21zs
45	BS 88-3	10.0/4.0	100, 102, C	58	NPzs	58	34zs
45	BS 88-3	16.0/6.0	100, 101, 102, A, C	91	NPzs	91	51zs

Notes to Table 7.1(i):

1 Voltage drop is the limiting constraint on the circuit cable length unless marked as follows:
  ▲ ad Limited by reduced csa of protective conductor (adiabatic limit)
  ▲ ol Cable/device/load combination not allowed in any of the installation conditions
  ▲ zs Limited by earth fault loop impedance Z_s
  ▲ sc Limited by line to neutral loop impedance (short-circuit).
2 The allowed installation methods are listed, see Tables 7.1(ii) and 7.1(iii) for further description.
3 NP - Not Permitted, prohibiting factor as note 1.
4 For application of RCDs and RCBOs, see 3.6.3.

▼ **Table 7.1(ii)** Installation reference methods and cable ratings for 70 °C thermoplastic (PVC) insulated and sheathed flat cable with protective conductor

Installation reference method		Conductor cross-sectional area (mm²)					
Ref.	Description	1.0	1.5	2.5	4	6	10	16
		A	A	A	A	A	A	A
C	Clipped direct	16	20	27	37	47	64	85
B*	Enclosed in conduit or trunking on a wall, etc.	13	16.5	23	30	38	52	69
102	In a stud wall with thermal insulation with cable touching the wall	13	16	21	27	35	47	63
100	In contact with plasterboard ceiling or joists covered by thermal insulation not exceeding 100 mm	13	16	21	27	34	45	57
A	Enclosed in conduit in an insulated wall	11.5	14.5	20	26	32	44	57
101	In contact with plasterboard ceiling or joists covered by thermal insulation exceeding 100 mm	10.5	13	17	22	27	36	46
103	Surrounded by thermal insulation including in a stud wall with thermal insulation with cable not touching a wall	8	10	13.5	17.5	23.5	32	42.5

Notes:

1 Cable ratings taken from Table 4D5 of BS 7671.
2 B* taken from Table 4D2A of BS 7671, see Appendix F.

▼ **Table 7.1(iii)** Installation methods specifically for flat twin and earth cables in thermal insulation

Installation Method			Reference Method to be used to determine current-carrying capacity
Number	Examples	Description	
100		Installation methods for flat twin and earth cable clipped direct to a wooden joist, or touching the plasterboard ceiling surface, above a plasterboard ceiling with thermal <u>insulation not exceeding</u> 100 mm in thickness having a minimum U value of 0.1 W/m²K	Table 4D5
101		Installation methods for flat twin and earth cable clipped direct to a wooden joist, or touching the plasterboard ceiling surface, above a plasterboard ceiling with thermal <u>insulation exceeding</u> 100 mm in thickness having a minimum U value of 0.1 W/m²K	Table 4D5
102		Installation methods for flat twin and earth cable in a stud wall with thermal insulation with a minimum U value of 0.1 W/m²K with the <u>cable touching</u> the inner wall surface, or touching the plasterboard ceiling surface, and the inner skin having a minimum U value of 10 W/m²K	Table 4D5
103		Installation methods for flat twin and earth cable in a stud wall with thermal insulation with a minimum U value of 0.1 W/m²K with the <u>cable not touching</u> the inner wall surface	Table 4D5

Notes:
1. Wherever practicable, a cable should be fixed in a position such that it will not be covered with thermal insulation.
2. Regulation 523.9, BS 5803-5: Appendix C 'Avoidance of overheating of electric cables', Building Regulations Approved Document B and Thermal Insulation: avoiding risks, BR 262, BRE 2001 refer.

7.2 Standard final circuits

7.2.1 Grouping of circuit cables

The tables assume heating (including water heating) cables are not grouped.

For cables of household or similar installations (heating and water heating excepted), if the following rules are followed, derating for grouping is not necessary:

    i   Cables are not grouped, that is, they are separated by at least two cable diameters when installed under thermal insulation, namely installation methods 100, 101, 102 and 103.
    ii   Cables clipped direct (including in cement or plaster) are clipped side by side in one layer and separated by at least one cable diameter.
    iii Cables above ceilings are clipped to joists as per installation methods 100 to 103 of Table 4A2 of BS 7671.

For other groupings, ambient temperatures higher than 30 °C or enclosure in thermal insulation, cable csa will need to be increased as per Appendix F of this Guide.

7.2.2 Socket-outlet circuits

The length represents the total ring cable loop length and does not include any spurs.

As a rule of thumb for rings, unfused spur lengths should not exceed 1/8 the cable length from the spur to the furthest point of the ring.

The total number of fused spurs is unlimited but the number of non-fused spurs is not to exceed the total number of socket-outlets and items of stationary equipment connected directly in the circuit.

A non-fused spur feeds only one twin socket-outlet or one permanently connected item of electrical equipment. Such a spur is connected to a circuit at the terminals of socket-outlets or at junction boxes or at the origin of the circuit in the distribution board.

A fused spur is connected to the circuit through a fused connection unit, the rating of the fuse in the unit not exceeding that of the cable forming the spur and, in any event, not exceeding 13 A. The number of socket-outlets which may be supplied by a fused spur is unlimited.

The circuit is assumed to have a load of 20 A at the furthest point and the balance to the rating of the protective device evenly distributed. (For a 32 A device this equates to a load of 26 A at the furthest point.)

7.2.3 Lighting circuits

A maximum voltage drop of 3 per cent of the 230 V nominal supply voltage has been allowed in the circuits; see Appendix F.

The circuit is assumed to have a load equal to the rated current (I_n) of the circuit protective device, evenly distributed along the circuit. Where this is not the case, circuit lengths will need to be reduced where voltage drop is the limiting factor, or halved where load is all at the extremity.

The most onerous installation condition acceptable for the load and device rating is presumed when calculating the limiting voltage drop. If the installation conditions are not the most onerous allowed (see column 4 of Table 7.1(i)) the voltage drop will not be as great as presumed in the table.

7.2.4 RCDs

Where circuits have residual current protection, the limiting factor is often the maximum loop impedance that will result in operation of the overcurrent device within 5 seconds for a short-circuit (line to neutral) fault. (See note 1 to Table 7.1(i) and limiting factor sc.)

7.2.5 Requirement for RCDs

RCDs are required:

411.5	i	where the earth fault loop impedance is too high to provide the required disconnection, for example, where the distributor does not provide a connection to the means of earthing – TT earthing arrangement
411.3.3(i)	ii	for socket-outlets where used by ordinary persons for general use
701.411.3.3	iii	for all circuits of locations containing a bath or shower
411.3.3(ii)	iv	for circuits supplying mobile equipment not exceeding 32 A for use outdoors
522.6.101	v	for cables without earthed metallic covering installed in walls or partitions at a depth of less than 50 mm and not protected by earthed steel conduit or similar
522.6.102		
522.6.103		
	vi	for cables without earthed metallic covering installed in walls or partitions with metal parts (not including screws or nails) and not protected by earthed steel conduit or the like.

RCD protection can be omitted in the following circumstances:

411.3.3(b)	i	specific labelled socket-outlets, for example, a socket-outlet for a freezer. However, the circuit cables must not require RCD protection as per v and vi above, that is, circuit cables must be enclosed in earthed steel conduit or have an earthed metal sheath or be at a depth of at least 50 mm in a wall or partition without metal parts
411.3.3(a)	ii	socket-outlet circuits in situations where the use of equipment and work on the building fabric and electrical installation is controlled by skilled or instructed persons, for example, in some industrial and commercial locations; see 3.6.2.2.

411.5.2	Cables installed on the surface do not specifically require RCD protection, however, RCD protection may be required for other reasons, such as where the installation forms part of a TT system and the earth fault loop impedance values for the overcurrent protective device cannot be met.

411.3.3(b) It is expected that all socket-outlets in a dwelling will have RCD protection at 30 mA, however, the exception of Regulation 411.3.3 can be applied in certain cases.

7.2.6 TT systems

For TT systems the figures for TN-C-S systems, with RCDs, may be used provided that:

    i   the circuit is protected by an RCD to BS 4293, BS EN 61008, BS EN 61009 or IEC 62325 with a rated residual operating current not exceeding that required for its circuit position,

    ii   the total earth fault loop impedance is verified as being less than 200 Ω, and

    iii a device giving both overload and short-circuit protection is installed in the circuit. This may be an RCBO or a combination of a fuse or circuit-breaker with an RCD.

7.2.7 Choice of protective device

The selection of protective device depends upon:

    i   prospective fault current
    ii   circuit load characteristics
    iii cable current-carrying capacity
    iv disconnection time limit.

Whilst these factors have generally been allowed for in the standard final circuits in Table 7.1(i), the following additional guidance is given:

    i   **Prospective fault current**

434.5.1   If a protective device is to operate safely, its rated short-circuit capacity must be not less than the prospective fault current at the point where it is installed. See Table 7.2.7(i).

313.1   The distributor needs to be consulted as to the prospective fault current at the origin of the installation. Except for London and some other major city centres, the maximum fault current for 230 V single-phase supplies up to 100 A will not exceed 16 kA. In general, the fault current is unlikely to exceed 16.5 kA.

▼ **Table 7.2.7(i)**   Rated short-circuit capacities

Device type	Device designation	Rated short-circuit capacity (kA)
Semi-enclosed fuse to BS 3036 with category of duty	S1A S2A S4A	1 2 4
Cartridge fuse to BS 1361 type I type II		16.5 33.0
General purpose fuse to BS 88-2		50 at 415 V
BS 88-3 type I type II		16 31.5
General purpose fuse to BS 88-6		16.5 at 240 V 80 at 415 V
Circuit-breakers to BS 3871 (replaced by BS EN 60898)	M1 M1.5 M3 M4.5 M6 M9	1 1.5 3 4.5 6 9
Circuit-breakers to BS EN 60898* and RCBOs to BS EN 61009		I_{cn} I_{cs} 1.5 (1.5) 3.0 (3.0) 6 (6.0) 10 (7.5) 15 (7.5) 20 (10.0) 25 (12.5)

* Two short-circuit capacities are defined in BS EN 60898 and BS EN 61009:

I_{cn} the rated short-circuit capacity (marked on the device).
I_{cs} the in-service short-circuit capacity.

The difference between the two is the condition of the circuit-breaker after manufacturer's testing.

I_{cn} is the maximum fault current the breaker can interrupt safely, although the breaker may no longer be usable.
I_{cs} is the maximum fault current the breaker can interrupt safely without loss of performance.

The I_{cn} value (in amperes) is normally marked on the device in a rectangle, for example, 6000 and for the majority of applications the prospective fault current at the terminals of the circuit-breaker should not exceed this value.

For domestic installations the prospective fault current is unlikely to exceed 6 kA, up to which value the I_{cn} will equal I_{cs}.

The short-circuit capacity of devices to BS EN 60947-2 is as specified by the manufacturer.

ii Circuit load characteristics

553.1.1.3

a *Semi-enclosed fuses.* Fuses should preferably be of the cartridge type. However, semi-enclosed fuses to BS 3036 are still permitted for use in domestic and similar premises if fitted with a fuse element which, in the absence of more specific advice from the manufacturer, meets the requirements of Table 53.1.

b *Cartridge fuses to BS 1361 (now withdrawn, replaced by BS 88-3:2010).* These are for use in domestic and similar premises.

c *Cartridge fuses to BS 88 series.* Three types are specified:
gG fuse links with a full-range breaking capacity for general application
gM fuse links with a full-range breaking capacity for the protection of motor circuits
aM fuse links for the protection of motor circuits.

d *Circuit-breakers to BS EN 60898 (or BS 3871-1) and RCBOs to BS EN 61009.* Guidance on selection is given in Table 7.2.7(ii).

▼ **Table 7.2.7(ii)** Application of circuit-breakers

Circuit-breaker type	Trip current (0.1 s to 5 s)	Application
1 B	2.7 to 4 I_n 3 to 5 I_n	Domestic and commercial installations having little or no switching surge
2 C 3	4 to 7 I_n 5 to 10 I_n 7 to 10 I_n	General use in commercial/industrial installations where the use of fluorescent lighting, small motors, etc., can produce switching surges that would operate a Type 1 or B circuit-breaker. Type C or 3 may be necessary in highly inductive circuits such as banks of fluorescent lighting
4 D	10 to 50 I_n 10 to 20 I_n	Not suitable for general use Suitable for transformers, X-ray machines, industrial welding equipment, etc., where high inrush currents may occur

NOTE: I_n is the nominal rating of the circuit-breaker.

iii Cable current-carrying capacities

For guidance on the coordination of device and cable ratings see Appendix F.

iv Disconnection times

411.3.2.2
411.3.2.3
411.3.2.4
411.8.3

The protective device must operate within the required disconnection time as appropriate for the circuit. Appendix B provides maximum permissible measured earth fault loop impedances for fuses, circuit-breakers and RCBOs.

7.3 Installation considerations

7.3.1 Floors and ceilings

522.6.100 Where a low voltage cable is installed under a floor or above a ceiling it must be run in such a position that it is not liable to be damaged by contact with the floor or ceiling or the fixings thereof. A cable passing through a joist or ceiling support must:

    i be at least 50 mm from the top or bottom, as appropriate, or
    ii have earthed armouring or an earthed metal sheath, or
    iii be enclosed in earthed steel conduit or trunking, or
    iv be provided with mechanical protection sufficient to prevent penetration of the cable by nails, screws and the like (NOTE: the requirement to prevent penetration is difficult to meet), or
414   **v** form part of a SELV or PELV circuit.

See Figure 7.3.1.

▼ **Figure 7.3.1** Cables through joists

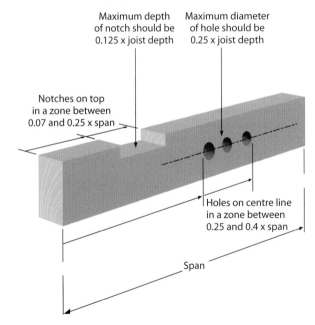

Notes:

1   Maximum diameter of hole should be 0.25 x joist depth.
2   Holes on centre line in a zone between 0.25 and 0.4 x span.
3   Maximum depth of notch should be 0.125 x joist depth.
4   Notches on top in a zone between 0.07 and 0.25 x span.
5   Holes in the same joist should be at least 3 diameters apart.

7.3.2 Walls and partitions

522.6.101 A cable concealed in a wall or partition must:

    i be at least 50 mm from the surface, or
    ii have earthed armouring or an earthed metal sheath, or
    iii be enclosed in earthed steel conduit or trunking, or
    iv be provided with mechanical protection sufficient to prevent penetration of the cable by nails, screws and the like (NOTE: the requirement to prevent penetration is difficult to meet), or
    v be installed either horizontally within 150 mm of the top of the wall or partition or vertically within 150 mm of the angle formed by two walls, or run horizontally or vertically to an accessory or consumer unit (see Figure 7.3.2), or

414   **vi** form part of a SELV or PELV circuit.

In domestic and similar installations, cables not installed as per i, ii, iii or iv but complying with v must be protected by a 30 mA RCD.

522.6.102 In domestic and similar installations, cables installed in walls or partitions with a metal or part metal construction must be either:

    a installed as ii, iii, iv or vi above, or
    b protected by a 30 mA RCD.

For installations under the supervision of a skilled or instructed person, such as commercial or industrial where only authorized equipment is used and only skilled persons will work on the building, RCD protection as described above is not required.

NOTE: Domestic or similar installations are not considered to be under the supervision of skilled or instructed persons.

▼ **Figure 7.3.2**   Zones prescribed in Regulation 522.6.101(v) (see v above)

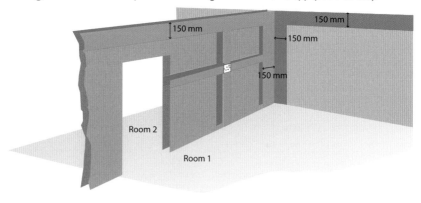

7.4 Proximity to electrical and other services

528.3 Electrical and all other services must be protected from any harmful mutual effects foreseen as likely under conditions of normal service. For example, cables should not be in contact with or run alongside hot pipes.

7.4.1 Segregation of Band I and Band II circuits

528.1 Part 2

Band I (extra-low voltage) circuits must not be contained within the same wiring system (for example, trunking) as Band II (low voltage) circuits unless:

  i every cable is insulated for the highest voltage present, or
  ii each conductor of a multicore cable is insulated for the highest voltage present, or
  iii the cables are installed in separate compartments, or
  iv the cables fixed to a cable tray are separated by a partition, or
  v for a multicore cable, they are separated by an earthed metal screen of equivalent current-carrying capacity to that of the largest Band II circuit.

Definitions of voltage bands

  ▶ Band I circuit: Circuit that is nominally extra-low voltage, i.e. not exceeding 50 V a.c. or 120 V d.c. For example, SELV, PELV, telecommunications, data and signalling
  ▶ Band II circuit: Circuit that is nominally low voltage, i.e. 51 to 1000 V a.c. and 121 to 1500 V d.c. Telecommunication cables that are generally ELV but have ringing voltages exceeding 50 V are Band I.

528.1, Note 2

NOTE: Fire alarm and emergency lighting circuits must be separated from other cables and from each other, in compliance with BS 5839 and BS 5266.

7.4.2 Proximity to communications cables

528.2 An adequate separation between telecommunication wiring (Band I) and electric power and lighting (Band II) circuits must be maintained. This is to prevent mains voltage appearing in telecommunication circuits with consequent danger to personnel. BS 6701:2004 recommends that the minimum separation distances given in Tables 7.4.2(i) and 7.4.2(ii) should be maintained.

▼ **Table 7.4.2(i)** External cables

Minimum separation distances between external low voltage electricity supply cables operating in excess of 50 V a.c. or 120 V d.c. to earth, but not exceeding 600 V a.c. or 900 V d.c. to earth (Band II), and telecommunications cables (Band I).

Voltage to earth	Normal separation distances	Exceptions to normal separation distances, plus conditions to exception
Exceeding 50 V a.c. or 120 V d.c., but not exceeding 600 V a.c. or 900 V d.c.	50 mm	Below this figure a non-conducting divider should be inserted between the cables

▼ **Table 7.4.2(ii)** Internal cables

Minimum separation distances between internal low voltage electricity supply cables operating in excess of 50 V a.c. or 120 V d.c. to earth, but not exceeding 600 V a.c. or 900 V d.c. to earth (Band II), and telecommunications cables (Band I).

Voltage to earth	Normal separation distances	Exceptions to normal separation distances, plus conditions to exception
Exceeding 50 V a.c. or 120 V d.c., but not exceeding 600 V a.c. or 900 V d.c.	50 mm	50 mm separation need not be maintained, provided that (i) the LV cables are enclosed in separate conduit which, if metallic, is earthed in accordance with BS 7671, OR (ii) the LV cables are enclosed in separate trunking which, if metallic, is earthed in accordance with BS 7671, OR (iii) the LV cable is of the mineral insulated type or is of earthed armoured construction.

Notes:

1 Where the LV cables share the same tray then the normal separation should be met.
2 Where LV and telecommunications cables are obliged to cross, additional insulation should be provided at the crossing point; this is not necessary if either cable is armoured.

7.4.3 Separation of gas installation pipework

Gas installation pipes must be spaced:

a at least 150 mm away from electricity meters, controls, electrical switches or socket-outlets, distribution boards or consumer units;
b at least 25 mm away from electricity supply and distribution cables.

See also 2.3 and Figure 2.3.

528.3.4 Note (The cited distances are quoted within BS 6891:2005(2008) *Installation of low pressure gas pipework in domestic premises*, clause 8.16.2.)

7.4.4 Induction loops

A particular form of harmful effect may occur when an electrical installation shares the space occupied by a hearing aid induction loop.

Under these circumstances, if line and neutral conductors or switch feeds and switch wires are not run close together, there may be interference with the induction loop.

This can occur when a conventional two-way lighting circuit is installed. This effect can be reduced by connecting as shown in Figure 7.4.4.

▼ **Figure 7.4.4** Circuit for reducing interference with induction loop

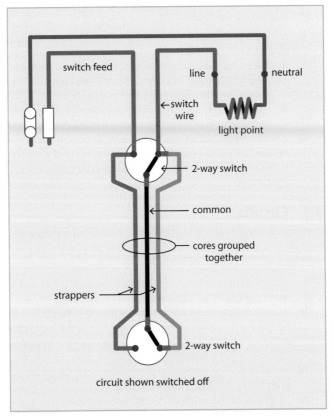

NOTE: Black/grey switch conductors to be identified in accordance with Table K1.

7.5 Earthing requirements for the installation of equipment having high protective conductor current

7.5.1 Equipment

Equipment having a protective conductor current exceeding 3.5 mA but not exceeding 10 mA must be either permanently connected to the fixed wiring of the installation or connected by means of an industrial plug and socket complying with BS EN 60309-2.

Equipment having a protective conductor current exceeding 10 mA should be connected by one of the following methods:

7

    i permanently connected to the wiring of the installation, with the protective conductor selected in accordance with Regulation 543.7.1.103. The permanent connection to the wiring may be by means of a flexible cable

    ii a flexible cable with an industrial plug and socket to BS EN 60309-2, provided that either:

       a the protective conductor of the associated flexible cable is of cross-sectional area not less than 2.5 mm² for plugs up to 16 A and not less than 4 mm² for plugs rated above 16 A, or

       b the protective conductor of the associated flexible cable is of cross-sectional area not less than that of the line conductor

    iii a protective conductor complying with Section 543 with an earth monitoring system to BS 4444 installed which, in the event of a continuity fault occurring in the protective conductor, automatically disconnects the supply to the equipment.

7.5.2 Circuits

543.7.1.103 The wiring of every final circuit and distribution circuit having a protective conductor current likely to exceed 10 mA must have high integrity protective conductor connections complying with one or more of the following:

    i a single protective conductor having a cross-sectional area not less than 10 mm², complying with Regulations 543.2 and 543.3

    ii a single copper protective conductor having a csa not less than 4 mm², complying with Regulations 543.2 and 543.3, the protective conductor being enclosed to provide additional protection against mechanical damage, for example within a flexible conduit

543.7.1.104   **iii** two individual protective conductors, each complying with Section 543, the ends being terminated independently

543.7.1.103   **iv** earth monitoring or use of double-wound transformer.

543.7.1.105 **NOTE:** Distribution boards are to indicate circuits with high protective conductor currents (see 6.15).

7.5.3 Socket-outlet final circuits

543.7.2.101 For a final circuit with socket-outlets or connection units, where the protective conductor current in normal service is likely to exceed 10 mA, the following arrangements are acceptable:

    i a ring final circuit with a ring protective conductor. Spurs, if provided, require high integrity protective conductor connections (Figure 7.5.3(i)), or

    ii a radial final circuit with:

       a a protective conductor connected as a ring (Figure 7.5.3(ii)), or

       b an additional protective conductor provided by metal conduit or ducting.

▼ **Figure 7.5.3(i)** Ring final circuit supplying socket-outlets

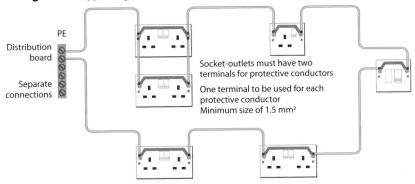

▼ **Figure 7.5.3(ii)** Radial final circuit supplying socket-outlets with duplicate protective conductors

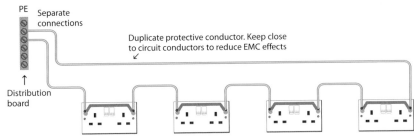

Socket-outlets must have two terminals for protective conductors
One terminal to be used for each protective conductor
Minimum size of 1.5 mm²

7.6 Electrical supplies to furniture

Where electrical equipment is installed within purpose-built items of furniture, such as cupboards, shop displays or lecterns, and supplied from a plug and socket arrangement, no specific standard exists for such installations, therefore guidance is given here which, essentially, follows the principles of BS 7671. For electrical systems in office furniture and educational furniture, BS 6396:2008 currently exists for installations which are supplied via a 13 A BS 1363 plug.

The following points should be adhered to:

415.1.1
- ▶ socket-outlets supplying items of furniture must be protected by an RCD providing additional protection at 30 mA
- ▶ cables of Band I and Band II circuits to be kept apart as far as is reasonably practicable; see also 7.4.1

7

- cables of Band I and Band II circuits, which are often hidden beneath the desk, should be sufficiently mechanically protected from damage caused by movement of chairs, storage of materials and the movement of feet and legs
- cable management systems or containment, such as conduit or trunking, should be installed to allow the safe routing, protection and separation of cables through the equipment
- long-term use of multi-gang extension leads should be avoided by installing a sufficient number of socket-outlets to supply the equipment to be used; employers should not allow ad hoc solutions to be created by users. See also see BS 6396:2008
- ensure that cables are sufficiently protected and cannot become trapped or damaged where desks are designed to be extended or altered to suit different activities or users.

543.2.1
543.2.6
There is no general requirement to ensure electrical continuity across the metallic frame of an item of furniture unless the frame has been designed to be used as a protective conductor.

Locations containing a bath or shower 8

8.1 Summary of requirements

701   Due to the presence of water, locations containing a bath or shower are onerous for equipment and there is an increased danger of electric shock.

The additional requirements can be summarised as follows:

701.411.3.3 **i**   all low voltage circuits of the location must be protected by 30 mA RCDs
701.512.3   **ii** socket-outlets, e.g. BS 1363, are not allowed within 3 metres of zone 1 (the edge of the bath or shower basin)
701.512.2   **iii** protection against ingress of water is specified for equipment within the zones, see Table 8.1 and Figures 8.1(i) to 8.1(iii)
701.512.3   **iv** there are restrictions as to where appliances, switchgear and wiring accessories may be installed, see Table 8.1 and Figures 8.1(i) to 8.1(iii).

701.415.2   Supplementary bonding of locations containing a bath or shower is required unless all the following requirements are met:

411.3.2.2   ▶ all circuits of the location meet the required disconnection times,
701.411.3.3 ▶ all circuits of the location have additional protection by 30 mA RCDs, and
411.3.1.2   ▶ all extraneous-conductive parts within the location are effectively connected by main protective bonding conductors to the main earthing terminal of the installation.

8

▼ **Table 8.1** Requirements for equipment (current-using and accessories) in a location containing a bath or shower

Zone	Minimum degree of protection	Current-using equipment	Switchgear and accessories
0	IPX7	Only 12 V a.c. rms or 30 V ripple-free d.c. SELV, the safety source installed outside the zones.	None allowed.
1	IPX4 (IPX5 if water jets)	25 V a.c. rms or 60 V ripple-free d.c. SELV or PELV, the safety source installed outside the zones. The following mains voltage fixed, permanently connected equipment allowed: whirlpool units, electric showers, shower pumps, ventilation equipment, towel rails, water heaters, luminaires.	Only 12 V a.c. rms or 30 V ripple-free d.c. SELV switches, the safety source installed outside the zones.
2	IPX4 (IPX5 if water jets)	Fixed permanently connected equipment allowed. General rules apply.	Only switches and sockets of SELV circuits allowed, the source being outside the zones, and shaver supply units complying with BS EN 61558-2-5 if fixed where direct spray is unlikely.
Outside zones	IPXXB or IP2X	General rules apply.	Accessories, SELV socket-outlets and shaver supply units to BS EN 61558-2-5 allowed. Socket-outlets allowed 3 m horizontally from the boundary of zone 1.

▼ **Figure 8.1(i)** Zone dimensions in a location containing a bath

Section

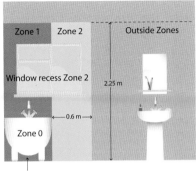

Plan

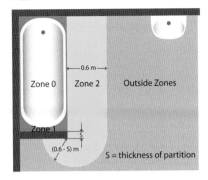

The space under the bath is:
Zone 1 if accessible without the use of a tool
outside the zones if accessible only with the use of a tool

▼ **Figure 8.1(ii)** Zones in a location containing a shower with basin and with permanent fixed partition

Section

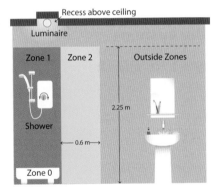

Plan

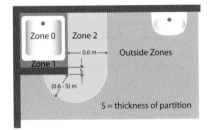

▼ **Figure 8.1(iii)** Zones in a location containing a shower without a basin, but with a partition

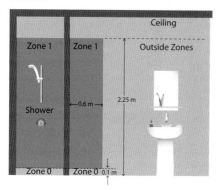
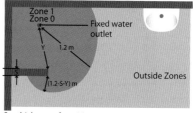

S = thickness of partition
Y = radial distance from the fixed water outlet to the inner corner of the partition

8.2 Shower cubicle in a room used for other purposes

Where a shower cubicle is installed in a room other than a bathroom or shower room the requirements for bathrooms and shower rooms must be complied with.

8.3 Underfloor heating systems

8.3.1 Locations containing a bath and shower

701.753 Underfloor heating installations in locations containing a bath and shower should have an overall earthed metallic grid or the heating cable should have an earthed metallic sheath, which must be connected to the protective conductor of the supply circuit.

8.3.2 Other areas

753.411.3.2 In areas other than special locations, Class I heating units which do not have an exposed-conductive-part, i.e. integrated earth screen or sheath, must have a metallic grid, with a spacing of not more than 30 mm, installed above the floor heating elements. The grid must be connected to the protective conductor of the electrical installation and the heating system protected by an RCD with a rated residual operating current not
415.1.1 exceeding 30 mA.

753.415.1 In areas where occupants are not expected to be completely wet, a circuit supplying
701.753 heating equipment of Class II construction or equivalent insulation should be provided with additional protection by the use of an RCD with a rated residual operating current not exceeding 30 mA.

Inspection and testing 9

9.1 Inspection and testing

610.1 Every installation must be inspected and tested during erection and on completion before being put into service to verify, so far as is reasonably practicable, that the requirements of the Regulations have been met.

Precautions must be taken to avoid danger to persons and to avoid damage to property and installed equipme nt during inspection and testing.

631.4
632.1
632.3
If the inspection and tests are satisfactory, a signed Electrical Installation Certificate together with a Schedule of Inspections and a Schedule of Test Results (as in Appendix G) are to be given to the person ordering the work.

9.2 Inspection

9.2.1 Procedure and purpose

611.1 Inspection must precede testing and must normally be done with that part of the installation under inspection disconnected from the supply.

611.2 The purpose of the inspection is to verify that equipment is:
  i   correctly selected and erected in accordance with BS 7671 (and, if appropriate, its own standard)
  ii not visibly damaged or defective so as to impair safety.

9.2.2 Inspection checklist

611.3 The inspection must include at least the checking of relevant items from the following checklist:

526	i	connection of conductors
514.3	ii	identification of conductors
522.6	iii	routing of cables in safe zones or protection against mechanical damage
433	iv	selection of conductors for current-carrying capacity and voltage drop, in
525		accordance with the design
132.14.1	v	connection of single-pole devices for protection or switching in line conductors only
526	vi	correct connection of accessories and equipment (including polarity)

On-Site Guide | **85**
© The Institution of Engineering and Technology

9

527.2	**vii** presence of fire barriers, suitable seals and protection against thermal effects
410.3.3	**viii** methods of protection against electric shock:

        a basic protection and fault protection, i.e.

414	▶ SELV
	▶ PELV
412	▶ double insulation
	▶ reinforced insulation

        b basic protection, i.e.

416.1	▶ insulation of live parts
416.2	▶ barriers or enclosures

        c fault protection

411	▶ automatic disconnection of supply

                The following to be confirmed for presence and sized in accordance with the design:
- earthing conductor
- circuit protective conductors
- protective bonding conductors
- earthing arrangements for combined protective and functional purposes
- presence of adequate arrangements for alternative source(s), where applicable
- FELV
- choice and setting of protective and monitoring devices (for fault and/or overcurrent protection)

413	▶ electrical separation
418.3	
415.1	**d** additional protection by RCDs
132.11	**ix** prevention of mutual detrimental influence (refer to 7.4)
537	**x** presence of appropriate devices for isolation and switching correctly located
445	**xi** presence of undervoltage protective devices (where appropriate)
514	**xii** labelling of protective devices including circuit-breakers, RCDs, fuses, switches and terminals, main earthing and bonding connections
522	**xiii** selection of equipment and protective measures appropriate to external influences
132.12	**xiv** adequacy of access to switchgear and equipment
514	**xv** presence of danger notices and other warning signs (see Section 6)
514.9	**xvi** presence of diagrams, instructions and similar information
522	**xvii** erection methods.

9.3 Testing

Testing must include the relevant tests from the following checklist.

612.1 When a test shows a failure to comply, the failure must be corrected. The test must then be repeated, as must any earlier test that could have been influenced by the failure.

9.3.1 Testing checklist

612.2   **i** continuity of conductors:

- protective conductors including main and supplementary bonding conductors
- ring final circuit conductors including protective conductors

612.3   **ii** insulation resistance (between live conductors and between each live conductor and earth). Where appropriate during this measurement, line and neutral conductors may be connected together, for example, where many lighting transformers are installed on a lighting circuit

612.6   **iii** polarity: this includes checks that single-pole control and protective devices, for example, switches, circuit-breakers and fuses, are connected in the line conductor only, that bayonet and Edison screw lampholders (except for E14 and E27 to BS EN 60238) have their outer contacts connected to the neutral conductor and that wiring has been correctly connected to socket-outlets and other accessories

612.7   **iv** earth electrode resistance (TT systems)

612.9   **v** earth fault loop impedance (TN systems)

612.11   **vi** prospective short-circuit current and prospective earth fault current, if not determined by enquiry of the distributor

612.13   **vii** functional testing, including:

- testing of RCDs
- operation of all switchgear

612.14   **viii** verification of voltage drop (not normally required during initial verification).

Guidance on initial testing of installations 10

10.1 Safety and equipment

HSR25, EWR Regulation 14

Electrical testing involves danger. The Electricity at Work Regulations 1989 state that working on live conductors is permissible provided that it is reasonable in all the circumstances for the work to be carried out and that suitable precautions are taken to prevent injury.

Live testing of electrical installations is, therefore, reasonable as it is a recognised method of assessing the suitability and safety of an electrical installation; suitable precautions must be taken by employing the correct test equipment and suitable personal protective equipment.

Although live testing and diagnosis for fault finding may be justifiable, there could be no justification for any subsequent repair work to be carried out live.

610.1
612.1

It is the test operative's duty to ensure their own safety, and the safety of others, whilst working through test procedures. When using test instruments, this is best achieved by precautions such as:

    i knowledge and experience of the correct application and use of the test instrumentation, leads, probes and accessories (is of the greatest importance)

612.1

    ii checking that the test instrumentation is made in accordance with the appropriate safety standards such as BS EN 61243-3 for two-pole voltage detectors and BS EN 61010 or BS EN 61557 for instruments

    iii checking before each use that all leads, probes, accessories (including all devices such as crocodile clips used to attach to conductors) and instruments including the proving unit are clean, undamaged and functioning; also, checking that isolation can be safely effected and that any locks or other means necessary for securing the isolation are available and functional

GS38

    iv observing the safety measures and procedures set out in HSE Guidance Note GS 38 for all instruments, leads, probes and accessories. Some test instrument manufacturers advise that their instruments be used in conjunction with fused test leads and probes. Others advise the use of non-fused leads and probes when the instrument has in-built electrical protection but it should be noted that such electrical protection does not extend to the probes and leads.

10.2 Sequence of tests

NOTE: The advice given does not preclude other test methods.

612.1 Tests should be carried out in the following sequence:

10.2.1 Before the supply is connected (i.e. isolated)

Ref		
612.2.1	**i**	continuity of protective conductors, including main and supplementary bonding
612.2.2	**ii**	continuity of ring final circuit conductors, including protective conductors
612.3	**iii**	insulation resistance
612.6	**iv**	polarity (by continuity method)
612.7	**v**	earth electrode resistance, using an earth electrode resistance tester (see vii also).

10.2.2 With the supply connected and energised

Ref		
GS38	**vi**	check polarity of supply, using an approved voltage indicator
612.7, Note	**vii**	earth electrode resistance, using a loop impedance tester
612.9	**viii**	earth fault loop impedance
612.11	**ix**	prospective fault current measurement, if not determined by enquiry of the distributor
612.13	**x**	functional testing, including RCDs and switchgear.

Results obtained during the various tests should be recorded on the Schedule of Test Results (Appendix G) for future reference and checked for acceptability against prescribed criteria.

10.3 Test procedures

612.2.1 10.3.1 Continuity of circuit protective conductors and protective bonding conductors (for ring final circuits see 10.3.2)

Test methods 1 and 2 are alternative ways of testing the continuity of protective conductors.

Every protective conductor, including circuit protective conductors, the earthing conductor, main and supplementary bonding conductors, should be tested to verify that the conductors are electrically sound and correctly connected.

Test method 1 detailed below, in addition to checking the continuity of the protective conductor, also measures ($R_1 + R_2$) which, when added to the external impedance (Ze), enables the earth fault loop impedance (Zs) to be checked against the design, see 10.3.6.

NOTE: ($R_1 + R_2$) is the sum of the resistances of the line conductor (R_1) and the circuit protective conductor (R_2) between the point of utilisation and origin of the installation.

Use an ohmmeter capable of measuring a low resistance for these tests.

Test method 1 can only be used to measure ($R_1 + R_2$) for an 'all-insulated' installation, such as an installation wired in 'twin and earth'. Installations incorporating steel conduit, steel trunking, MICC and PVC/SWA cables will produce parallel paths to protective conductors. Such installations should be inspected for soundness of construction and test method 1 or 2 used to prove continuity.

612.2.1 **i   Continuity of circuit protective conductors**

Continuity test method 1

Bridge the line conductor to the protective conductor at the distribution board so as to include all the circuit. Then test between line and earth terminals at each point in the circuit. The measurement at the circuit's extremity should be recorded and is the value of ($R_1 + R_2$) for the circuit under test (see Figure 10.3.1(i)).

If the instrument does not include an 'auto-null' facility, or this is not used, the resistance of the test leads should be measured and deducted from the resistance readings obtained.

▼ **Figure 10.3.1(i)**   Connections for testing continuity of circuit protective conductors using test method 1

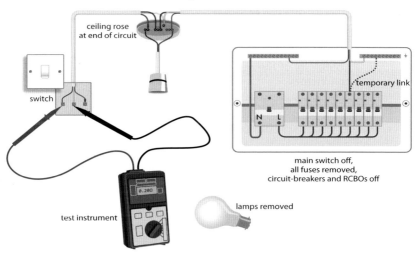

Continuity test method 2

Connect one terminal of the test instrument to a long test lead and connect this to the installation main earthing terminal.

Connect the other terminal of the instrument to another test lead and use this to make contact with the protective conductor at various points on the circuit, such as luminaires, switches, spur outlets, etc. (see Figure 10.3.1(ii)).

If the instrument does not include an 'auto-null' facility, or this is not used, the resistance of the test leads should be measured and deducted from the resistance readings obtained.

The resistance of the protective conductor R_2 is recorded on the Schedule of Test Results; see Appendix G.

10

▼ **Figure 10.3.1(ii)** Continuity test method 2

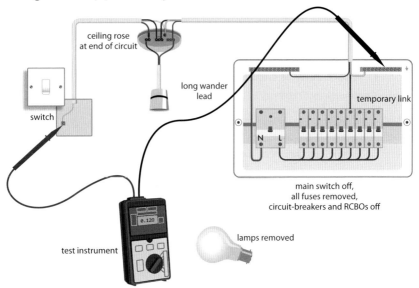

ii Continuity of the earthing conductor and protective bonding conductors

Continuity test method 2
For main bonding, connect one terminal of the test instrument to a long test lead and connect this to the installation main earthing terminal. Connect the other terminal of the instrument to another test lead and use this to make contact with the protective bonding conductor at its further end, such as at its connection to the incoming metal water, gas or oil service.

The *Continuity and connection verified* boxes on the Electrical Installation Certificate should be ticked if the continuity and connection of the earthing conductor and of each main bonding conductor are satisfactory. The details of the material and the cross-sectional areas of the conductors must also be recorded.

10.3.2 Continuity of ring final circuit conductors

A three-step test is required to verify the continuity of the line, neutral and protective conductors and the correct wiring of a ring final circuit. The test results show if the ring has been interconnected to create an apparently continuous ring circuit which is in fact broken, or wrongly wired.

Use a low-resistance ohmmeter for this test.

Step 1

The line, neutral and protective conductors are identified at the distribution board and the end-to-end resistance of each is measured separately (see Figure 10.3.2(i)). These resistances are r_1, r_n and r_2 respectively. A finite reading confirms that there is no open circuit on the ring conductors under test. The resistance values obtained should be the same (within 0.05 Ω) if the conductors are all of the same size. If the protective conductor has a reduced csa the resistance r_2 of the protective conductor loop will be proportionally higher than that of the line and neutral loops, for example, 1.67 times for 2.5/1.5 mm² cable. If these relationships are not achieved then either the conductors are incorrectly identified or there is something wrong at one or more of the accessories.

▼ **Figure 10.3.2(i)** Step 1: The end-to-end resistances of the line, neutral and protective conductors are measured separately

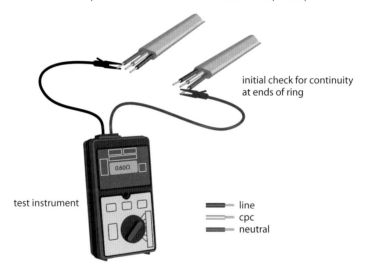

Step 2

The line and neutral conductors are then connected together at the distribution board so that the outgoing line conductor is connected to the returning neutral conductor and vice versa (see Figure 10.3.2(ii)). The resistance between line and neutral conductors is measured at each socket-outlet. The readings at each of the socket-outlets wired into the ring will be substantially the same and the value will be approximately one-quarter of the resistance of the line plus the neutral loop resistances, i.e. $(r_1 + r_n)/4$. Any socket-outlets wired as spurs will have a higher resistance value due to the resistance of the spur conductors.

NOTE: Where single-core cables are used, care should be taken to verify that the line and neutral conductors of *opposite ends* of the ring circuit are connected together. An error in this respect will be apparent from the readings taken at the socket-outlets, progressively increasing in value as readings are taken towards the midpoint of the ring, then decreasing again towards the other end of the ring.

▼ **Figure 10.3.2(ii)** Step 2: The line and neutral conductors are cross-connected and the resistance measured at each socket-outlet

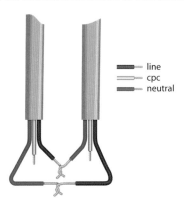

Step 3

The above step is then repeated, this time with the line and cpc cross-connected at the distribution board (see Figure 10.3.2(iii)). The resistance between line and earth is measured at each socket-outlet. The readings obtained at each of the socket-outlets wired into the ring will be substantially the same and the value will be approximately one-quarter of the resistance of the line plus cpc loop resistances, i.e. $(r_1 + r_2)/4$. As before, a higher resistance value will be measured at any socket-outlets wired as spurs. The highest value recorded represents the maximum $(R_1 + R_2)$ of the circuit and is recorded on the Schedule of Test Results. The value can be used to determine the earth fault loop impedance (Z_s) of the circuit to verify compliance with the loop impedance requirements of BS 7671 (see 10.3.6).

▼ **Figure 10.3.2(iii)** Step 3: The line conductors and cpc are cross-connected and the resistance measured at each socket-outlet

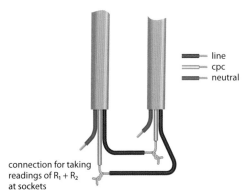

This sequence of tests also verifies the polarity of each socket-outlet, except that where the testing has been carried out at the terminals on the reverse of the accessories, a

visual inspection is required to confirm correct polarity connections, and dispenses with the need for a separate polarity test.

10.3.3 Insulation resistance

i Pre-test checks

    a Pilot or indicator lamps and capacitors are disconnected from circuits to prevent misleading test values from being obtained

    b If a circuit includes voltage-sensitive electronic devices such as RCCBs, RCBOs or SRCDs incorporating electronic amplifiers, dimmer switches, touch switches, delay timers, power controllers, electronic starters or controlgear for fluorescent lamps, etc., either:

        1 the devices must be temporarily disconnected, or

        2 a measurement should be made between the live conductors (line and neutral) connected together and the protective earth only.

ii Tests

Tests should be carried out using the appropriate d.c. test voltage specified in Table 10.3.3.

The tests should be made at the distribution board or consumer unit with the main switch off.

When testing simple installations, i.e. those consisting of one consumer unit only, the installation could be tested as a whole with all fuses in place, switches and circuit-breakers closed, lamps removed and other current-using equipment disconnected; see Figure 10.3.3(i).

▼ **Figure 10.3.3(i)** Insulation resistance test of the whole installation

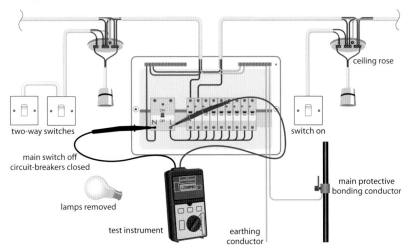

When testing individual circuits, it is important to remove the fuse or open the circuit-breaker of that circuit; this ensures that no other circuits at the board influence the result of the test.

Where the removal of lamps and/or the disconnection of current-using equipment is impracticable, the local switches controlling such lamps and/or equipment should be open.

Where a circuit contains two-way switching, the two-way switches must be operated one at a time and further insulation resistance tests carried out to ensure that all the circuit wiring is tested.

▼ Table 10.3.3 Minimum values of insulation resistance

Table 61

Circuit nominal voltage	Test voltage (V d.c.)	Minimum insulation resistance (MΩ)
SELV and PELV	250	0.5
Up to and including 500 V with the exception of SELV and PELV, but including FELV	500	1.0

Notes:
1. Insulation resistance measurements are usually much higher than those of Table 10.3.3.
2. More stringent requirements are applicable for the wiring of fire alarm systems in buildings; see BS 5839-1.

For an installation operating at 400/230 V, although an insulation resistance value of only 1 MΩ complies with BS 7671, where the insulation resistance measured is less than 2 MΩ the possibility of a latent defect exists. In these circumstances, each circuit should then be tested separately.

Where surge protective devices (SPDs) or other equipment such as electronic devices or RCDs with amplifiers are likely to influence the results of the test or may suffer damage from the test voltage, such equipment must be disconnected before carrying out the insulation resistance test.

612.3.2 Where it is not reasonably practicable to disconnect such equipment, the test voltage for the particular circuit may be reduced to 250 V d.c. but the insulation resistance must be at least 1 MΩ.

Where the circuit includes electronic devices which are likely to influence the results or be damaged, only a measurement between the live conductors connected together and earth should be made and the reading should be not less than the value stated in Table 10.3.3.

iii Insulation resistance between live conductors

Single-phase and three-phase
Test between all the live (line and neutral) conductors at the distribution board (see Figure 10.3.3(i)).

Figure 10.3.3(ii) shows an insulation resistance test performed between live conductors of a single circuit.

Resistance readings obtained should be not less than the value stated in Table 10.3.3.

▼ **Figure 10.3.3(ii)** Insulation resistance test between live conductors of a circuit

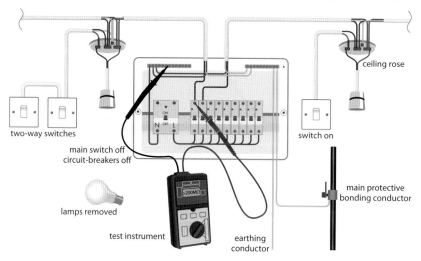

NOTE: The test may initially be carried out on the complete installation.

iv Insulation resistance to earth

Single-phase

Test between the live conductors (line and neutral) and the circuit protective conductors at the distribution board (Figure 10.3.3(iii) illustrates neutral to earth only).

For a circuit containing two-way switching or two-way and intermediate switching, the switches must be operated one at a time and the circuit subjected to additional insulation resistance tests.

▼ **Figure 10.3.3(iii)** Insulation resistance test between neutral and earth

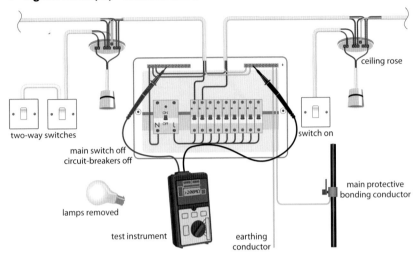

Notes:

1. The test may initially be carried out on the complete installation.
2. Earthing and bonding connections are in place.
3. The earthing conductor must connect the main earthing terminal to the means of earthing whilst testing.

612.3.1

Three-phase

Test to earth from all live conductors (including the neutral) connected together. Where a low reading is obtained it is necessary to test each conductor separately to earth, after disconnecting all equipment.

Resistance readings obtained should be not less than the value stated in Table 10.3.3.

v SELV and PELV circuits

612.4.1 Test between SELV and PELV circuits and live parts of other circuits at 500 V d.c.

612.4.2

Test between SELV or PELV conductors at 250 V d.c. and between PELV conductors and protective conductors of the PELV circuit at 250 V d.c.

Resistance readings obtained should be not less than the value stated in Table 10.3.3.

vi FELV circuits

612.4.4 FELV circuits are tested as low voltage circuits at 500 V d.c.

10.3.4 Polarity

See Figure 10.3.4.

The method of test prior to connecting the supply is the same as test method 1 for checking the continuity of protective conductors which should have already been carried out (see 10.3.1). For ring final circuits a visual check may be required (see 10.3.2 following step 3).

It is important to confirm that:

1. overcurrent devices and single-pole controls are in the line conductor,
2. except for E14 and E27 lampholders to BS EN 60238, centre contact screw lampholders have the outer threaded contact connected to the neutral, and
3. socket-outlet and similar accessory polarities are correct.

After connection of the supply, correct polarity must be confirmed using a voltage indicator or a test lamp (in either case with leads complying with the recommendations of HSE Guidance Note GS 38).

GS 38

▼ **Figure 10.3.4** Polarity test on a lighting circuit

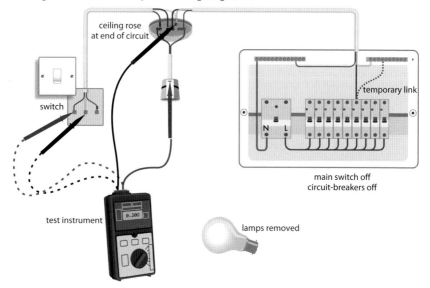

NOTE: The test may be carried out either at lighting points or switches.

10.3.5 Earth electrode resistance measurement

10.3.5.1 Loop impedance method

If the electrode under test is being used in conjunction with an RCD protecting an installation forming part of a TT system, the following method of test may be applied.

A loop impedance tester is connected between the line conductor at the origin of the installation and the earth electrode with the test link open and a test performed. This impedance reading is treated as the electrode resistance and is then added to the resistance of the protective conductor for the protected circuits. The test should be carried out before energising the remainder of the installation.

Table 41.5 Note 2
The measured resistance should meet the following criteria and those of 10.3.6 but, in any event, should not exceed 200 Ω.

411.5.3
For TT systems, the value of the earth electrode resistance R_A in ohms multiplied by the operating current in amperes of the protective device $I_{\Delta n}$ should not exceed 50 V.

For example, if $R_A = 200$ Ω, then the maximum RCD operating current should not exceed 250 mA.

REMEMBER TO REPLACE THE TEST LINK.

10.3.5.2 Proprietary earth electrode test instrument

The test requires the use of two temporary test spikes (electrodes), and is carried out in the following manner.

Connection to the earth electrode, E, is made using terminals C1 and P1 of a four-terminal earth tester. To exclude the resistance of the test leads from the resistance reading, individual leads should be taken from these terminals and connected separately to the electrode. If the test lead resistance is insignificant, the two terminals may be short-circuited at the tester and connection made with a single test lead, the same being true if using a three-terminal tester. Connection to the temporary spikes is made as shown in Figure 10.3.5.2. The distance between the test spikes is important. If they are too close together, their resistance areas will overlap.

In general, reliable results may be expected if the distance between the electrode under test and the current spike T1 is at least ten times the maximum dimension of the electrode system, for example, 30 m for a 3 m long rod electrode. With an auxiliary electrode T2 inserted halfway between the electrode under test E and temporary electrode T1, the voltage drop between E and T2 is measured. The resistance of the electrode is then obtained by the test instrument from the voltage between E and T2 divided by the current flowing between E and T1, provided that there is no overlap of the resistance areas.

To confirm that the electrode resistance obtained above is a true value, two further readings are taken, firstly with electrode T2 moved ≈6 m further from electrode E and secondly with electrode T2 moved 6 m closer to electrode E. If the results obtained from the three tests above are substantially the same, the average of the three readings is taken as the resistance of the earth electrode under test. If the results obtained are significantly different, the above procedure should be repeated with test electrode T1 placed further from the electrode under test.

▼ **Figure 10.3.5.2** Earth electrode test

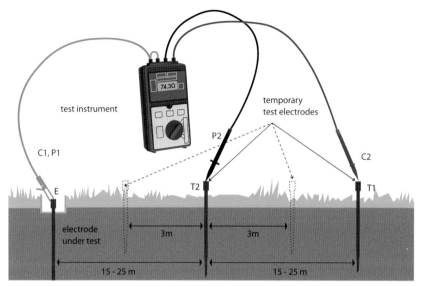

The instrument output current may be a.c. or reversed d.c. to overcome electrolytic effects. As these types of test instrument employ phase-sensitive detectors (PSD), the errors associated with stray currents are eliminated. The instrument should be capable of checking that the resistance of the temporary spikes used for testing is within the accuracy limits stated in the instrument specification. This may be achieved by an indicator provided on the instrument, or the instrument should have a sufficiently high upper range to enable a discrete test to be performed on the spikes. If the temporary spike resistance is too high, measures to reduce the resistance will be necessary, such as driving the spikes deeper into the ground.

10.3.6 Earth fault loop impedance

612.9 The earth fault loop impedance (Z_s) is required to be determined for the furthest point of each circuit. It may be determined by:

- ▶ direct measurement of Zs, or
- ▶ direct measurement of Ze at the origin and adding ($R_1 + R_2$) measured during the continuity tests (10.3.1 and 10.3.2) {$Zs = Ze + (R_1 + R_2)$}, or
- ▶ adding ($R_1 + R_2$) measured during the continuity tests to the value of Ze declared by the distributor (see 1.1(iv) and 1.3(iv)).

The effectiveness of the distributor's earth must be confirmed by a test.

The external impedance (Z_e) may be measured using a line-earth loop impedance tester.

The main switch is opened and made secure to isolate the installation from the source of supply. The earthing conductor is disconnected from the main earthing terminal and the measurement made between line and earth of the supply.

REMEMBER TO RECONNECT THE EARTHING CONDUCTOR TO THE EARTH TERMINAL AFTER THE TEST.

Direct measurement of Z_s can only be made on a live installation. Neither the connection with earth nor bonding conductors are disconnected. The reading given by the loop impedance tester will usually be less than $Z_e + (R_1 + R_2)$ because of parallel earth return paths provided by any bonded extraneous-conductive-parts. This must be taken into account when comparing the results with design data.

610.1 Care should be taken to avoid any shock hazard to the testing personnel and to other persons on site during the tests.

The value of Z_s determined for each circuit should not exceed the value given in Appendix B for the particular overcurrent device and cable.

411.4.9 For TN systems, when protection is afforded by an RCD, the rated residual operating current in amperes times the earth fault loop impedance in ohms should not exceed 50 V. This test should be carried out before energising other parts of the system.

NOTE: For further information on the measurement of earth fault loop impedance, refer to IET Guidance Note 3 — *Inspection & Testing*.

10.3.7 Measurement of prospective fault current
612.11

It is not recommended that installation designs are based on measured values of prospective fault current, as changes to the distribution network subsequent to completion of the installation may increase fault levels.

Designs should be based on the maximum fault current provided by the distributor (see 7.2.7(i)).

If it is desired to measure prospective fault levels this should be done with all main bonding in place. Measurements are made at the distribution board between live conductors and between line conductors and earth.

For three-phase supplies, the maximum possible fault level will be approximately twice the single-phase to neutral value. (For three-phase to earth faults, neutral and earth path impedances have no influence.)

10.3.8 Check of phase sequence
612.12

In the case of three-phase circuits, it should be verified that the phase sequence is maintained.

10.3.9 Functional testing

612.13 RCDs should be tested as described in Section 11.

Switchgear, controls, etc., should be functionally tested; that is, operated to check that they work and are properly mounted and installed.

10.3.10 Verification of voltage drop

NOTE: Verification of voltage drop is not normally required during initial verification.

612.14 Where required, it should be verified that voltage drop does not exceed the limits stated in relevant product standards of installed equipment.

525.100 Where no such limits are stated, voltage drop should be such that it does not impair the proper and safe functioning of installed equipment.

Typically, voltage drop will be evaluated using the measured circuit impedances.

The requirements for voltage drop are deemed to be met where the voltage drop between the origin and the relevant piece of equipment does not exceed the values stated in Appendix 4 of BS 7671:2008(2011).

Appx. 4
Table 4Ab
Appendix 4, paragraph 6.4, gives maximum values of voltage drop for both lighting and other uses and depending upon whether the installation is supplied directly from an LV distribution system or from a private LV supply.

It should be remembered that voltage drop may exceed the values stated in Appendix 4 in situations, such as motor starting periods and where equipment has a high inrush current, where such events remain within the limits specified in the relevant product standard or reasonable recommendation by an equipment manufacturer.

Operation of RCDs   11

Residual current device (RCD) is the generic term for a device that operates when the residual current in the circuit reaches a predetermined value. An RCD is a protective device used to automatically disconnect the electrical supply when an imbalance is detected between the line and neutral conductors. In the case of a single-phase circuit, see Figure 11.0, the device monitors the difference in currents between the line and neutral conductors. In a healthy circuit, where there is no earth fault current or protective conductor current, the sum of the currents in the line and neutral conductors is zero. If a line to earth fault develops, a portion of the line conductor current will not return through the neutral conductor. The device monitors this difference, operates and disconnects the circuit when the residual current reaches a preset limit, the residual operating current ($I_{\Delta n}$).

▼ **Figure 11.0** RCD operation

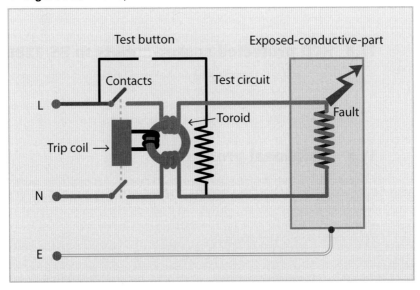

11

11.1 General test procedure

612.8

The tests are made on the load side of the RCD, as near as practicable to its point of installation and between the line conductor of the protected circuit and the associated circuit protective conductor. The load supplied should be disconnected during the test.

11.2 General-purpose RCCBs to BS 4293

i With a leakage current flowing equivalent to 50 per cent of the rated tripping current of the RCD, the device should not open.

ii With a leakage current flowing equivalent to 100 per cent of the rated tripping current of the RCD, the device should open in less than 200 ms. Where the RCD incorporates an intentional time delay it should trip within a time range from '50% of the rated time delay plus 200 ms' to '100% of the rated time delay plus 200 ms'.

11.3 General-purpose RCCBs to BS EN 61008 or RCBOs to BS EN 61009

i With a leakage current flowing equivalent to 50 per cent of the rated tripping current of the RCD, the device should not open.

ii With a leakage current flowing equivalent to 100 per cent of the rated tripping current of the RCD, the device should open in less than 300 ms unless it is of 'Type S' (or selective) which incorporates an intentional time delay. In this case, it should trip within a time range from 130 ms to 500 ms.

11.4 RCD protected socket-outlets to BS 7288

i With a leakage current flowing equivalent to 50 per cent of the rated tripping current of the RCD, the device should not open.

ii With a leakage current flowing equivalent to 100 per cent of the rated tripping current of the RCD, the device should open in less than 200 ms.

11.5 Additional protection

612.10
415.1.1

Where an RCD with a rated residual operating current $I_{\Delta n}$ not exceeding 30 mA is used to provide additional protection (against direct contact), with a test current of 5 $I_{\Delta n}$ the device should open in less than 40 ms. The maximum test time must not be longer than 40 ms, unless the protective conductor potential rises by less than 50 V. (The instrument supplier will advise on compliance.)

11.6 Integral test device

612.13.1 An integral test device is incorporated in each RCD. This device enables the electrical and mechanical parts of the RCD to be verified, by pressing the button marked 'T' or 'Test' (Figure 11.0).

Operation of the integral test device does **not** provide a means of checking:

- **a** the continuity of the earthing conductor or the associated circuit protective conductors
- **b** any earth electrode or other means of earthing
- **c** any other part of the associated installation earthing.

The test button will only operate the RCD if the device is energised.

Confirm that the notice to test RCDs quarterly (by pressing the test button) is fixed in a prominent position (see 6.12).

11.7 Multipole RCDs

As each live conductor of the RCD is incorporated in the magnetic sensing circuit it is not necessary to perform the test for poles L2 and L3. However, if there is any doubt as to the authenticity of the device in question - in terms of a fake or counterfeit device – the advice would be to repeat the test for poles L2 and L3. It goes without saying that such important devices, designed to protect life and property, should be obtained from trusted sources and made by reputable manufacturers.

If a decision is made to test the RCD on all three lines, there should be little on no discernable difference in operating times as each pole is incorporated in the magnetic sensing circuit. If, for example, the test performed on one pole did not meet the required disconnection time, yet tests on the other two poles were satisfactory, the device should be considered faulty and replaced.

Appendix A
Maximum demand and diversity

This appendix provides information on the determination of the maximum demand for an installation and includes the current demand to be assumed for commonly used equipment. It also includes some notes on the application of allowances for diversity.

The information and values given in this appendix are intended only for guidance because it is impossible to specify the appropriate allowances for diversity for every type of installation and such allowances call for special knowledge and experience. The values given in Table A2, therefore, may be increased or decreased as decided by the installation designer concerned. No guidance is given for blocks of residential dwellings, large hotels, industrial and large commercial premises; such installations should be assessed on a case-by-case basis.

The current demand of a final circuit is determined by adding the current demands of all points of utilisation and equipment in the circuit and, where appropriate, making an allowance for diversity. Typical current demands to be used for this addition are given in Table A1.

The current demand of an installation consisting of a number of final circuits may be assessed by using the allowances for diversity given in Table A2 which are applied to the total current demand of all the equipment supplied by the installation. The current demand of the installation should not be assessed by adding the current demands of the individual final circuits obtained as outlined above. In Table A2 the allowances are expressed either as percentages of the current demand or, where followed by the letters f.l. (full load), as percentages of the rated full load current of the current-using equipment. The current demand for any final circuit which is a standard circuit arrangement complying with Appendix H is the rated current of the overcurrent protective device of that circuit.

An alternative method of assessing the current demand of an installation supplying a number of final circuits is to add the diversified current demands of the individual circuits and then apply a further allowance for diversity. In this method the allowances given in Table A2 should not be used, the values to be chosen being the responsibility of the installation designer.

A | Appendix

The use of other methods of determining maximum demand is not precluded where specified by the installation designer. After the design currents for all the circuits have been determined, enabling the conductor sizes to be chosen, it is necessary to check that the limitation on voltage drop is met.

▼ **Table A1** Current demand to be assumed for points of utilisation and current-using equipment

Point of utilisation or current-using equipment	Current demand to be assumed
Socket-outlets other than 2 A socket-outlets and other than 13 A socket-outlets See note 1	Rated current
2 A socket-outlets	At least 0.5 A
Lighting outlet See note 2	Current equivalent to the connected load, with a minimum of 100 W per lampholder
Electric clock, shaver supply unit (complying with BS EN 61558-2-5), shaver socket-outlet (complying with BS 4573), bell transformer, and current-using equipment of a rating not greater than 5 VA	May be neglected for the purpose of this assessment
Household cooking appliance	The first 10 A of the rated current plus 30% of the remainder of the rated current plus 5 A if a socket-outlet is incorporated in the control unit
All other stationary equipment	British Standard rated current, or normal current

Notes:

1. See Appendix H for the design of standard circuits using socket-outlets to BS 1363-2 and BS EN 60309-2 (BS 4343).
2. Final circuits for discharge lighting must be arranged so as to be capable of carrying the total steady current, viz. that of the lamp(s) and any associated controlgear and also their harmonic currents. Where more exact information is not available, the demand in volt-amperes is taken as the rated lamp watts multiplied by not less than 1.8. This multiplier is based upon the assumption that the circuit is corrected to a power factor of not less than 0.85 lagging, and takes into account controlgear losses and harmonic current.

Appendix A

▼ **Table A2** Allowances for diversity (see opposite for notes * and †)

Purpose of the final circuit fed from the conductors or switchgear to which the diversity applies	Individual household installations including individual dwellings of a block	Type of premises — Small shops, stores, offices and business premises	Type of premises — Small hotels, boarding houses, guest houses, etc.
1 Lighting	66% of total current demand	90% of total current demand	75% of total current demand
2 Heating and power (but see 3 to 8 below)	100% of total current demand up to 10 a +50% of any current demand in excess of 10 a	100% f.l. of largest appliance +75% f.l. of remaining appliances	100% f.l. of largest appliance +80% f.l. of second largest appliance +60% f.l. of remaining appliances
3 Cooking appliances	10 a + 30% f.l. of connected cooking appliances in excess of 10 a + 5 a if a socket-outlet is incorporated in the control unit	100% f.l. of largest appliance +80% f.l. of second largest appliance +60% f.l. of remaining appliances	100% f.l. of largest appliance +80% f.l. of second largest appliance +60% f.l. of remaining appliances
4 Motors (other than lift motors, which are subject to special consideration)	Not applicable	100% f.l. of largest motor +80% f.l. of second largest motor +60% f.l. of remaining motors	100% f.l. of largest motor +50% f.l. of remaining motors
5 Water-heaters (instantaneous type)*	100% f.l. of largest appliance +100% f.l. of second largest appliance +25% f.l. of remaining appliances	100% f.l. of largest appliance +100% f.l. of second largest appliance +25% f.l. of remaining appliances	100% f.l. of largest appliance +100% f.l. of second largest appliance +25% f.l. of remaining appliances
6 Water-heaters (thermostatically controlled)	No diversity allowable†		
7 Floor warming installations	No diversity allowable†		
8 Thermal storage space heating installations	No diversity allowable†		

Appendix

▼ **Table A2** *continued*

Purpose of the final circuit fed from the conductors or switchgear to which the diversity applies	Type of premises		
	Individual household installations including individual dwellings of a block	Small shops, stores, offices and business premises	Small hotels, boarding houses, guest houses, etc.
9 Standard arrangement of final circuits in accordance with Appendix H	100% of current demand of largest circuit +40% of current demand of every other circuit	100% of current demand of largest circuit +50% of current demand of every other circuit	
10 Socket-outlets (other than those included in 9 above and stationary equipment other than those listed above)	100% of current demand of largest point of utilisation +40% of current demand of every other point of utilisation	100% of current demand of largest point of utilisation +70% of current demand of every other point of utilisation	100% of current demand of largest point of utilisation +75% of current demand of every other point in main rooms (dining rooms, etc.) +40% of current demand of every other point of utilisation

Notes to Table A2:

* In this context an instantaneous water-heater is considered to be a water-heater of any loading which heats water only while the tap is turned on and therefore uses electricity intermittently.

† It is important to ensure that distribution boards or consumer units are of sufficient rating to take the total load connected to them without the application of any diversity.

Appendix B
Maximum permissible measured earth fault loop impedance

612.9
411.4.6
411.4.7
411.4.8

The tables in this appendix provide maximum permissible measured earth fault loop impedances (Zs) for compliance with BS 7671 where the standard final circuits of Table 7.1(i) are used. The values are those that must not be exceeded in the tests carried out under 10.3.6 at an ambient temperature of 10 °C. Table B8 provides correction factors for other ambient temperatures.

Where the cables to be used are to Table 4, 7 or 8 of BS 6004 or Table 3, 5, 6 or 7 of BS 7211 or are other thermoplastic (PVC) or thermosetting (low smoke halogen-free – LSHF) cables to these British Standards and the cable loading is such that the maximum operating temperature is 70 °C, then Tables B1–B5 give the maximum earth fault loop impedances for circuits with:

1. protective conductors of copper and having from 1 mm^2 to 16 mm^2 cross-sectional area
2. an overcurrent protective device (i.e. a fuse) to:
   - BS 3036 (Table B1)
   - BS 88-2.2 and BS 88-6 (Table B2)
   - BS 88-2 (Table B3)
   - BS 88-3 (Table B4)
   - BS 1361 (Table B5).

For each type of fuse, two tables are given:

411.3.2.2
▶ where the circuit concerned is a final circuit not exceeding 32 A and the maximum disconnection time for compliance with Regulation 411.3.2.2 is 0.4 s for TN systems, and

411.3.2.3
▶ where the circuit concerned is a final circuit exceeding 32 A or a distribution circuit, and the disconnection time for compliance with Regulation 411.3.2.3 is 5 s for TN systems.

543.1.3
In each table the earth fault loop impedances given correspond to the appropriate disconnection time from a comparison of the time/current characteristics of the device concerned and the equation given in Regulation 543.1.3.

B Appendix

The tabulated values apply only when the nominal voltage to Earth (U_0) is 230 V.

Table B6 gives the maximum measured Zs for circuits protected by circuit-breakers to BS 3871-1 and BS EN 60898 and RCBOs to BS EN 61009.

Note: The impedances tabulated in this appendix are lower than those in Tables 41.2, 41.3 and 41.4 of BS 7671 as the impedances in this appendix are measured values at an assumed conductor temperature of 10 °C whilst those in BS 7671 are design figures at the conductor normal operating temperature. The correction factor (divisor) used is 1.24. For smaller section cables the impedance may also be limited by the adiabatic equation of Regulation 543.1.3. A value of k of 115 from Table 54.3 of BS 7671 is used. This is suitable for PVC insulated and sheathed cables to Table 4, 7 or 8 of BS 6004 and for thermosetting (LSHF) insulated and sheathed cables to Table 3, 5, 6 or 7 of BS 7211. The k value is based on both the thermoplastic (PVC) and LSHF cables operating at a maximum temperature of 70 °C.

The IET *Commentary on the Wiring Regulations* provides more information.

▼ **Table B1** Semi-enclosed fuses. Maximum measured earth fault loop impedance (in ohms) at ambient temperature where the overcurrent protective device is a semi-enclosed fuse to BS 3036

i 0.4 second disconnection (final circuits not exceeding 32 A in TN systems)

Protective conductor (mm²)	Fuse rating (A)			
	5	15	20	30
1.0	7.7	2.1	1.4	NP
≥ 1.5	7.7	2.1	1.4	0.9

ii 5 seconds disconnection (final circuits exceeding 32 A and distribution circuits in TN systems)

Protective conductor (mm²)	Fuse rating (A)			
	20	30	45	60
1.0	2.7	NP	NP	NP
1.5	3.1	2.0	NP	NP
2.5	3.1	2.1	1.2	NP
4.0	3.1	2.1	1.3	0.8
≥ 6.0	3.1	2.1	1.3	0.9

NOTE: NP means that the combination of the protective conductor and the fuse is Not Permitted.

Appendix B

▼ **Table B2** BS 88-2.2 and BS 88-6 fuses. Maximum measured earth fault loop impedance (in ohms) at ambient temperature where the overcurrent protective device is a fuse to BS 88-2.2 or BS 88-6

i 0.4 second disconnection (final circuits not exceeding 32 A in TN systems)

Protective conductor (mm²)	Fuse rating (A)					
	6	10	16	20	25	32
1.0	6.9	4.1	2.2	1.4	1.2	0.66
1.5	6.9	4.1	2.2	1.4	1.2	0.84
≥ 2.5	6.9	4.1	2.2	1.4	1.2	0.84

ii 5 seconds disconnection (final circuits exceeding 32 A and distribution circuits in TN systems)

Protective conductor (mm²)	Fuse rating (A)							
	20	25	32	40	50	63	80	100
1.0	1.7	1.2	0.66	NP	NP	NP	NP	NP
1.5	2.3	1.7	1.1	0.64	NP	NP	NP	NP
2.5	2.3	1.8	1.5	0.93	0.55	0.34	NP	NP
4.0	2.3	1.8	1.5	1.1	0.77	0.50	0.23	NP
6.0	2.3	1.8	1.5	1.1	0.84	0.66	0.36	0.22
10.0	2.3	1.8	1.5	1.1	0.84	0.66	0.46	0.33
16.0	2.3	1.8	1.5	1.1	0.84	0.66	0.46	0.34

NOTE: NP means that the combination of the protective conductor and the fuse is Not Permitted.

B Appendix

▼ **Table B3** BS 88-2 fuses. Maximum measured earth fault loop impedance (in ohms) at ambient temperature where the overcurrent protective device is a fuse to BS 88-2

i 0.4 second disconnection (final circuits not exceeding 32 A in TN systems)

Protective conductor (mm²)	Fuse rating (A)					
	6	10	16	20	25	32
1.0	6.6	3.9	2.0	1.4	1.1	0.63ad
1.5	6.6	3.9	2.0	1.4	1.1	0.83
≥ 2.5	6.6	3.9	2.0	1.4	1.1	0.83

ii 5 seconds disconnection (final circuits exceeding 32 A and distribution circuits in TN systems)

Protective conductor (mm²)	Fuse rating (A)							
	20	25	32	40	50	63	80	100
1.0	1.67ad	1.02ad	0.64ad	NP	NP	NP	NP	NP
1.5	2.36	1.31ad	0.97ad	0.57ad	NP	NP	NP	NP
2.5	2.36	1.8	1.47	0.89ad	0.55ad	0.33ad	NP	NP
4.0	2.36	1.8	1.47	1.1	0.75ad	0.53ad	0.25ad	NP
6.0	2.36	1.8	1.47	1.1	0.84	0.66	0.36ad	0.22ad
10.0	2.36	1.8	1.47	1.1	0.84	0.66	0.46	0.32ad
16.0	2.36	1.8	1.47	1.1	0.84	0.66	0.46	0.37

NOTE 1: NP means that the combination of the protective conductor and the fuse is Not Permitted.
NOTE 2: ad – adiabatic limitation.

Appendix B

▼ **Table B4** BS 88-3 fuses. Maximum measured earth fault loop impedance (in ohms) at ambient temperature where the overcurrent protective device is a semi-enclosed fuse to BS 88-3

i 0.4 second disconnection (final circuits not exceeding 32 A in TN systems)

Protective conductor (mm²)	Fuse rating (A)			
	5	16	20	32
1.0	8.36	1.94	1.63	0.6ad
1.5 to 16	8.36	1.94	1.63	0.77

ii 5 seconds disconnection (final circuits exceeding 32 A and distribution circuits in TN systems)

Protective conductor (mm²)	Fuse rating (A)					
	20	32	45	63	80	100
1.0	2.32ad	0.62ad	NP	NP	NP	NP
1.5	2.72	0.84ad	NP	NP	NP	NP
2.5	2.72	1.23ad	0.62ad	0.25ad	NP	NP
4.0	2.72	1.32	0.84	0.41ad	0.23ad	0.13ad
6.0	2.72	1.32	0.84	0.58	0.33ad	0.19ad
10	2.72	1.32	0.84	0.58	0.43	0.32
16	2.72	1.32	0.84	0.58	0.43	0.32

NOTE 1: NP means that the combination of the protective conductor and the fuse is Not Permitted.
NOTE 2: ad – adiabatic limitation.

B Appendix

▼ **Table B5** BS 1361 fuses. Maximum measured earth fault loop impedance (in ohms) at ambient temperature where the overcurrent protective device is a semi-enclosed fuse to BS 1361

i 0.4 second disconnection (final circuits not exceeding 32 A in TN systems)

Protective conductor (mm²)	Fuse rating (A)			
	5	15	20	30
1.0	8.4	2.6	1.4	0.81
1.5	8.4	2.6	1.4	0.93
2.5 to 16	8.4	2.62	1.4	0.93

ii 5 seconds disconnection (final circuits exceeding 32 A and distribution circuits in TN systems)

Protective conductor (mm²)	Fuse rating (A)					
	20	30	45	60	80	100
1.0	1.7	0.81	NP	NP	NP	NP
1.5	2.2	1.2	0.34	NP	NP	NP
2.5	2.3	1.5	0.52	0.21	NP	NP
4.0	2.3	1.5	0.69	0.37	0.22	NP
6.0	2.3	1.5	0.77	0.53	0.30	0.15
10	2.3	1.5	0.77	0.56	0.40	0.22
16	2.3	1.5	0.77	0.56	0.40	0.29

NOTE: NP means that the combination of the protective conductor and the fuse is Not Permitted.

Appendix **B**

▼ **Table B6** Circuit-breakers. Maximum measured earth fault loop impedance (in ohms) at ambient temperature where the overcurrent device is a circuit-breaker to BS 3871 or BS EN 60898 or RCBO to BS EN 61009

0.1 to 5 second disconnection times

Circuit-breaker type	Circuit-breaker rating (A)												
	5	6	10	15	16	20	25	30	32	40	45	50	63	100
1	9.27	7.73	4.64	3.09	2.90	2.32	1.85	1.55	1.45	1.16	1.03	0.93	0.74	0.46
2	5.3	4.42	2.65	1.77	1.66	1.32	1.06	0.88	0.83	0.66	0.59	0.53	0.42	0.26
B	7.42	6.18	3.71	2.47	2.32	1.85	1.48	1.24	1.16	0.93	0.82	0.74	0.59	0.37
3&C	3.71	3.09	1.85	1.24	1.16	0.93	0.74	0.62	0.58	0.46	0.41	0.37	0.29	0.19
D	1.85	1.55	0.93	0.62	0.58	0.46	0.37	0.31	0.29	0.23	0.21	0.19	0.15	0.09

Regulation 434.5.2 of BS 7671:2008(2011) requires that the protective conductor csa meets the requirements of BS EN 60898-1, -2 or BS EN 61009-1, or the minimum quoted by the manufacturer. The sizes given in Table B7 are for energy limiting class 3, Types B and C devices only.

B | Appendix

▼ **Table B7** Minimum protective conductor size (mm²)*

Energy limiting class 3 device rating	Fault level (kA)	Protective conductor csa (mm²)	
		Type B	Type C
Up to and including 16 A	≤ 3	1.0	1.5
Up to and including 16 A	≤ 6	2.5	2.5
Over 16 up to and including 32 A	≤ 3	1.5	1.5
Over 16 up to and including 32 A	≤ 6	2.5	2.5
40 A	≤ 3	1.5	1.5
40 A	≤ 6	2.5	2.5

* For other device types and ratings or higher fault levels, consult manufacturer's data. See Regulation 434.5.2 and the IET publication *Commentary on the IEE Wiring Regulations*.

▼ **Table B8** Ambient temperature correction factors

Ambient temperature (°C)	Correction factor (from 10 °C) (notes 1 and 2)
0	0.96
5	0.98
10	1.00
20	1.04
25	1.06
30	1.08

Notes:
1. The correction factor is given by: $\{1 + 0.004(\text{ambient temp} - 10)\}$ where 0.004 is the simplified resistance coefficient per °C at 20 °C given by BS EN 60228 for both copper and aluminium conductors.
2. The factors are different to those of Table I2 because Table B8 corrects from 10 °C and Table I2 from 20 °C.

The appropriate ambient correction factor from Table B8 is applied to the earth fault loop impedances of Tables B1–B6 if the ambient temperature is other than 10 °C when the circuit loop impedances are measured.

For example, if the ambient temperature is 25 °C the measured earth fault loop impedance of a circuit protected by a 32 A type B circuit-breaker to BS EN 60898 should not exceed 1.16 x 1.06 = 1.23 Ω.

Appendix C
Selection of types of cable for particular uses and external influences

52 For compliance with the requirements of Chapter 52 for the selection and erection of wiring systems in relation to risks of mechanical damage and corrosion, this appendix lists, in two tables, types of cable for the uses indicated. These tables are not intended to be exhaustive and other limitations may be imposed by the relevant regulations of BS 7671, in particular, those concerning maximum permissible operating temperatures.

Information is also included in this appendix on protection against corrosion of exposed metalwork of wiring systems.

C Appendix

▼ **Table C1** Applications of cables for fixed wiring

Type of cable (note 7)	Uses	Comments
Thermoplastic (PVC) or thermosetting insulated non-sheathed cable (BS 7211, BS 7919)	For use in conduits, cable ducting or trunking	Intermediate support may be required on long vertical runs 70 °C maximum conductor temperature for normal wiring grades including thermosetting types (note 4) Cables run in PVC conduit should not operate with a conductor temperature greater than 70 °C (note 4)
Flat thermoplastic (PVC) or thermosetting insulated and sheathed cable (BS 6004)	For general indoor use in dry or damp locations. May be embedded in plaster For use on exterior surface walls, boundary walls and the like For use as overhead wiring between buildings For use underground in conduits or pipes For use in building voids or ducts formed in-situ	Additional mechanical protection may be necessary where exposed to mechanical stresses Protection from direct sunlight may be necessary. Black sheath colour is better for cables exposed to sunlight May need to be hard drawn (HD) copper conductors for overhead wiring (note 6) Unsuitable for embedding directly in concrete
Mineral insulated (BS EN 60702-1)	General	MI cables should have overall PVC covering where exposed to the weather or risk of corrosion, or where installed underground, or in concrete ducts
Thermoplastic or thermosetting insulated, armoured, thermoplastic sheathed (BS 5467, BS 6346, BS 6724, BS 7846)	General	Additional protection may be necessary where exposed to mechanical stresses Protection from direct sunlight may be necessary. Black sheath colour is better for cables exposed to sunlight

Notes:

1. The use of cable covers or equivalent mechanical protection is desirable for all underground cables which might otherwise subsequently be disturbed. Route marker tape should also be installed, buried just below ground level. Cables should be buried at a sufficient depth.
2. Cables having thermoplastic (PVC) insulation or sheath should preferably not be used where the ambient temperature is consistently below 0 °C or has been within the preceding 24 hours. Where they are to be installed during a period of low temperature, precautions should be taken to avoid

Appendix C

risk of mechanical damage during handling. A minimum ambient temperature of 5 °C is advised in BS 7540:2005 (series) *Electric cables – Guide to use for cables with a rated voltage not exceeding 450/750 V* for some types of PVC insulated and sheathed cables.

3 Cables must be suitable for the maximum ambient temperature, and must be protected from any excess heat produced by other equipment, including other cables.

4 Thermosetting cable types (to BS 7211 or BS 5467) can operate with a conductor temperature of 90 °C. This must be limited to 70 °C where drawn into a conduit, etc., with thermoplastic (PVC) insulated conductors or connected to electrical equipment (512.1.5 and 523.1), or where such cables are installed in plastic conduit or trunking.

5 For cables to BS 6004, BS 6007, BS 7211, BS 6346, BS 5467 and BS 6724, further guidance may be obtained from those standards. Additional advice is given in BS 7540:2005 (series) *Guide to use of cables with a rated voltage not exceeding 450/750 V* for cables to BS 6004, BS 6007 and BS 7211.

6 Cables for overhead wiring between buildings must be able to support their own weight and any imposed wind or ice/snow loading. A catenary support is usual but hard drawn copper types may be used.

7 **BS 5467: Electric cables.** Thermosetting insulated, armoured cables for voltages of 600/1000 V and 1900/3300 V

BS 6004: Electric cables. PVC insulated, non-armoured cables for voltages up to and including 450/750 V for electric power, lighting and internal wiring

BS 6346: Electric cables. PVC insulated, armoured cables for voltages of 600/1000 V and 1900/3300 V

BS 6724: Electric cables. Thermosetting insulated, armoured cables for voltages of 600/1000 V and 1900/3300 V, having low emission of smoke and corrosive gases when affected by fire

BS 7211: Electric cables. Thermosetting insulated, non-armoured cables for voltages up to and including 450/750 V, for electric power, lighting and internal wiring, and having low emission of smoke and corrosive gases when affected by fire

BS 7846: Electric cables. 600/1000 V armoured fire-resistant cables having thermosetting insulation and low emission of smoke and corrosive gases when affected by fire

BS EN 60702-1: Mineral insulated cables and their terminations with a rated voltage not exceeding 750 V. Cables

Migration of plasticiser from thermoplastic (PVC) materials

Thermoplastic (PVC) sheathed cables, including thermosetting insulated with thermoplastic sheath, e.g. LSHF, must be separated from expanded polystyrene materials to prevent take-up of the cable plasticiser by the polystyrene as this will reduce the flexibility of the cables.

Thermal insulation

Thermoplastic (PVC) sheathed cables in roof spaces must be clipped clear of any insulation made of expanded polystyrene granules.

Cable clips

Polystyrene cable clips are softened by contact with thermoplastic (PVC). Nylon and polypropylene are unaffected.

Grommets

Natural rubber grommets can be softened by contact with thermoplastic (PVC). Synthetic rubbers are more resistant. Thermoplastic (PVC) grommets are not affected, but could affect other plastics.

Wood preservatives

Thermoplastic (PVC) sheathed cables should be covered to prevent contact with preservative fluids during application. After the solvent has evaporated (good ventilation is necessary) the preservative has no effect.

Appendix C

Creosote

Creosote should not be applied to thermoplastic (PVC) sheathed cables because it causes decomposition, solution, swelling and loss of pliability.

▼ **Table C2** Applications of flexible cables to BS 6500:2000 and BS 7919:2001

Type of flexible cable	Uses
Light thermoplastic (PVC) insulated and sheathed flexible cable	Indoors in household or commercial premises in dry situations, for light duty
Ordinary thermoplastic (PVC) insulated and sheathed flexible cable	Indoors in household or commercial premises, including damp situations, for medium duty For cooking and heating appliances where not in contact with hot parts For outdoor use other than in agricultural or industrial applications For electrically powered hand tools
60 °C thermosetting (rubber) insulated braided twin and three-core flexible cable	Indoors in household or commercial premises where subject only to low mechanical stresses
60 °C thermosetting (rubber) insulated and sheathed flexible cable	Indoors in household or commercial premises where subject only to low mechanical stresses For occasional use outdoors For electrically powered hand tools
60 °C thermosetting (rubber) insulated oil-resisting with flame-retardant sheath	For general use, unless subject to severe mechanical stresses For use in fixed installations where protected by conduit or other enclosure
90 °C thermosetting (rubber) insulated HOFR sheathed	General, including hot situations, e.g. night storage heaters, immersion heaters and boilers
90 °C heat-resisting thermoplastic (PVC) insulated and sheathed	General, including hot situations, e.g. for pendant luminaires
150 °C thermosetting (rubber) insulated and braided	For use at high ambient temperatures For use in or on luminaires
185 °C glass-fibre insulated single-core, twisted twin and three-core	For internal wiring of luminaires only and then only where permitted by BS 4533
185 °C glass-fibre insulated braided circular	For dry situations at high ambient temperatures and not subject to abrasion or undue flexing For the wiring of luminaires

Notes:
1. Cables having thermoplastic (PVC) insulation or sheath should preferably not be used where the ambient temperature is consistently below 0 °C. Where they are to be installed during a period of low temperature, precautions should be taken to avoid risk of mechanical damage during handling.
2. Cables should be suitable for the maximum ambient temperature, and should be protected from any excess heat produced by other equipment, including other cables.

Appendix C

3. For flexible cords and cables to BS 6007, BS 6141 and BS 6500 further guidance may be obtained from those standards, or from BS 7540:2005 (series) *Guide to use of cables with a rated voltage not exceeding 450/750 V.*
4. Where used as connections to equipment, flexible cables should, where possible, be of the minimum practicable length to minimize danger. The length of the flexible cable must be such that will permit correct operation of the protective device.
5. Where attached to equipment flexible cables should be protected against tension, crushing, abrasion, torsion and kinking, particularly at the inlet point to the electrical equipment. At such inlet points it may be necessary to use a device which ensures that the cable is not bent to an internal radius below that given in the appropriate part of Table 4 of BS 6700. Strain relief, clamping devices or cable guards should not damage the cable.
6. Flexible cables should not be run under carpets or other floor coverings where furniture or other equipment may rest on them or where heat dissipation from the cable will be affected. Flexible cables should not be placed where there is a risk of damage from traffic passing over them, unless suitably protected.
7. Flexible cables should not be used in contact with or close to heated surfaces, especially if the surface approaches the upper thermal limit of the cable.

Protection of exposed metalwork and wiring systems against corrosion

522.3
522.5

In damp situations, where metal cable sheaths and armour of cables, metal conduit and conduit fittings, metal ducting and trunking systems, and associated metal fixings, are liable to chemical deterioration or electrolytic attack by materials of a structure with which they may come in contact, it is necessary to take suitable precautions against corrosion.

Materials likely to cause such attack include:

- ▶ materials containing magnesium chloride which are used in the construction of floors and plaster mouldings
- ▶ plaster undercoats which may include corrosive salts
- ▶ lime, cement and plaster, for example on unpainted walls
- ▶ oak and other acidic woods
- ▶ dissimilar metals likely to set up electrolytic action.

Application of suitable coatings before erection or prevention of contact by separation with plastics, are recognized as effective precautions against corrosion.

Special care is required in the choice of materials for clips and other fittings for bare aluminium sheathed cables and for aluminium conduit, to avoid risk of local corrosion in damp situations. Examples of suitable materials for this purpose are the following:

- ▶ porcelain
- ▶ plastics
- ▶ aluminium
- ▶ corrosion-resistant aluminium alloys
- ▶ zinc alloys complying with BS 1004
- ▶ iron or steel protected against corrosion by galvanizing, sherardizing, etc.

522.5.2 Contact between bare aluminium sheaths or aluminium conduits and any parts made of brass or other metal having a high copper content should be especially avoided in damp situations, unless the parts are suitably plated. If such contact is unavoidable, the joint should be completely protected against ingress of moisture. Wiped joints in aluminium sheathed cables should always be protected against moisture by a suitable paint, by an impervious tape, or by embedding in bitumen.

Appendix D
Methods of support for cables, conductors and wiring systems

522.8　This appendix describes examples of methods of support for cables, conductors and wiring systems which should satisfy the relevant requirements of Chapter 52 of BS 7671. The use of other methods is not precluded where specified by a suitably qualified electrical engineer.

Cables generally

Items 1 to 8 below are generally applicable to supports on structures which are subject only to vibration of low severity and a low risk of mechanical impact.

1. For non-sheathed cables, installation in conduit without further fixing of the cables, precautions being taken against undue compression or other mechanical stressing of the insulation at the top of any vertical runs exceeding 5 m in length.
2. For cables of any type, installation in ducting or trunking without further fixing of the cables, vertical runs not exceeding 5 m in length without intermediate support.
3. For sheathed and/or armoured cables installed in accessible positions, support by clips at spacings not exceeding the appropriate value stated in Table D1.
4. For cables of any type, resting without fixing in horizontal runs of ducts, conduits, cable ducting or trunking.
5. For sheathed and/or armoured cables in horizontal runs which are inaccessible and unlikely to be disturbed, resting without fixing on part of a building, the surface of that part being reasonably smooth.
6. For sheathed-and-armoured cables in vertical runs which are inaccessible and unlikely to be disturbed, supported at the top of the run by a clip and a rounded support of a radius not less than the appropriate value stated in Table D5.
7. For sheathed cables without armour in vertical runs which are inaccessible and unlikely to be disturbed, supported by the method described in Item 6 above; the length of run without intermediate support not exceeding 5 m for a thermosetting or thermoplastic sheathed cable.
8. For thermosetting or thermoplastic (PVC) sheathed cables, installation in conduit without further fixing of the cables, any vertical runs being in conduit of suitable size and not exceeding 5 m in length.

D | Appendix

Particular applications

721.522.8
9 In caravans, for sheathed cables in inaccessible spaces such as ceiling, wall and floor spaces, support at intervals not exceeding 0.4 m for vertical runs and 0.25 m for horizontal runs.

10 In caravans, for horizontal runs of sheathed cables passing through floor or ceiling joists in inaccessible floor or ceiling spaces, securely bedded in thermal insulating material, no further fixing is required.

11 For flexible cables used as pendants, attachment to a ceiling rose or similar accessory by the cable grip or other method of strain relief provided in the accessory.

12 For temporary installations and installations on construction sites, supports so arranged that there is no appreciable mechanical strain on any cable termination or joint.

Overhead wiring

13 For cables sheathed with thermosetting or thermoplastic material, supported by a separate catenary wire, either continuously bound up with the cable or attached thereto at intervals, the intervals not exceeding those stated in column 2 of Table D1.

14 Support by a catenary wire incorporated in the cable during manufacture, the spacings between supports not exceeding those stated by the manufacturer and the minimum height above ground being in accordance with Table D2.

15 For spans without intermediate support (e.g. between buildings) of thermoplastic (PVC) insulated thermoplastic (PVC) sheathed cable, or thermosetting insulated cable having an oil-resisting and flame-retardant or HOFR sheath, terminal supports so arranged that:

- ▶ no undue strain is placed upon the conductors or insulation of the cable,
- ▶ adequate precautions are taken against any risk of chafing of the cable sheath, and
- ▶ the minimum height above ground and the length of such spans are in accordance with the appropriate values indicated in Table D2.

16 Bare or thermoplastic (PVC) covered conductors of an overhead line for distribution between a building and a remote point of utilisation (e.g. another building) supported on insulators, the lengths of span and heights above ground having the appropriate values indicated in Table D2 or otherwise installed in accordance with the Electricity Safety, Quality and Continuity Regulations 2002 (as amended).

17 For spans without intermediate support (e.g. between buildings) and which are in situations inaccessible to vehicular traffic, cables installed in heavy gauge steel conduit, the length of span and height above ground being in accordance with Table D2.

Conduit and cable trunking

18 Rigid conduit supported in accordance with Table D3.
19 Cable trunking supported in accordance with Table D4.
20 Conduit embedded in the material of the building.
21 Pliable conduit embedded in the material of the building or in the ground, or supported in accordance with Table D3.

Appendix D

▼ **Table D1** Spacings of supports for cables in accessible positions

Overall diameter of cable, d* (mm)	Maximum spacings of clips (mm)							
	Non-armoured thermosetting or thermoplastic (PVC) sheathed cables				Armoured cables		Mineral insulated copper sheathed or aluminium sheathed cables	
	Generally		In caravans					
	Horizontal †	Vertical †	Horizontal †	Vertical †	Horizontal †	Vertical †	Horizontal †	Vertical †
1	2	3	4	5	6	7	8	9
d ≤ 9	250	400	250 (for all sizes)	400 (for all sizes)	–	–	600	800
9 < d ≤ 15	300	400			350	450	900	1200
15 < d ≤ 20	350	450			400	550	1500	2000
20 < d ≤ 40	400	550			450	600	–	–

Note: For the spacing of supports for cables having an overall diameter exceeding 40 mm, the manufacturer's recommendations should be observed.
* For flat cables taken as the dimension of the major axis.
† The spacings stated for horizontal runs may be applied also to runs at an angle of more than 30° from the vertical. For runs at an angle of 30° or less from the vertical, the vertical spacings are applicable.

D Appendix

▼ **Table D2** Maximum lengths of span and minimum heights above ground for overhead wiring between buildings, etc.

Type of system	Maximum length of span (m)	Minimum height of span above ground (m)†		
		At road crossings	In positions accessible to vehicular traffic, other than crossings	In positions inaccessible to vehicular traffic*
1	2	3	4	5
Cables sheathed with thermoplastic (PVC) or having an oil-resisting and flame-retardant or HOFR sheath, without intermediate support.	3	5.8	5.8	3.5
Cables sheathed with thermoplastic (PVC) or having an oil-resisting and flame-retardant or HOFR sheath, in heavy gauge steel conduit of diameter not less than 20 mm and not jointed in its span.	3	5.8	5.8	3
Thermoplastic (PVC) covered overhead lines on insulators without intermediate support.	30	5.8	5.8	3.5
Bare overhead lines on insulators without intermediate support.	30	5.8	5.8	5.2
Cables sheathed with thermoplastic (PVC) or having an oil-resisting and flame-retardant or HOFR sheath, supported by a catenary wire.	No limit	5.8	5.8	3.5
Aerial cables incorporating a catenary wire.	Subject to Item 14	5.8	5.8	3.5
A bare or insulated overhead line for distribution between buildings and structures must be installed to the standard required by the Electricity Safety, Quality and Continuity Regulations 2002.				

* Column 5 is not applicable in agricultural premises.
† In some special cases, such as where cranes are present, it will be necessary to increase the minimum height of span above ground. It is preferable to use underground cables in such locations.

Appendix D

▼ **Table D3** Spacings of supports for conduits

Nominal diameter of conduit, d (mm)	Maximum distance between supports (m)					
	Rigid metal		Rigid insulating		Pliable	
	Horizontal	Vertical	Horizontal	Vertical	Horizontal	Vertical
1	2	3	4	5	6	7
$d \leq 16$	0.75	1.0	0.75	1.0	0.3	0.5
$16 < d \leq 25$	1.75	2.0	1.5	1.75	0.4	0.6
$25 < d \leq 40$	2.0	2.25	1.75	2.0	0.6	0.8
$d > 40$	2.25	2.5	2.0	2.0	0.8	1.0

Notes:
1. The spacings tabulated allow for maximum fill of cables permitted by the Regulations and the thermal limits specified in the relevant British Standards. They assume that the conduit is not exposed to other mechanical stress.
2. Supports should be positioned within 300 mm of bends or fittings. A flexible conduit should be of such length that it does not need to be supported in its run.
3. The inner radius of a conduit bend should be not less than 2.5 times the outside diameter of the conduit.

▼ **Table D4** Spacings of supports for cable trunking

Cross-sectional area of trunking, A (mm²)	Maximum distance between supports (m)			
	Metal		Insulating	
	Horizontal	Vertical	Horizontal	Vertical
1	2	3	4	5
$300 < A \leq 700$	0.75	1.0	0.5	0.5
$700 < A \leq 1500$	1.25	1.5	0.5	0.5
$1500 < A \leq 2500$	1.75	2.0	1.25	1.25
$2500 < A \leq 5000$	3.0	3.0	1.5	2.0
$A > 5000$	3.0	3.0	1.75	2.0

Notes:
1. The spacings tabulated allow for maximum fill of cables permitted by the Regulations and the thermal limits specified in the relevant British Standards. They assume that the trunking is not exposed to other mechanical stress.
2. The above figures do not apply to lighting suspension trunking, where the manufacturer's instructions must be followed, or where special strengthening couplers are used. Supports should be positioned within 300 mm of bends or fittings.

D Appendix

▼ **Table D5** Minimum internal radii of bends in cables for fixed wiring

Insulation	Finish	Overall diameter, d* (mm)	Factor to be applied to overall diameter of cable to determine minimum internal radius of bend
Thermosetting or thermoplastic (PVC) (circular, or circular stranded copper or aluminium conductors)	Non-armoured	$d \leq 10$	3(2)†
		$10 < d \leq 25$	4(3)†
		$d > 25$	6
	Armoured	Any	6
Thermosetting or thermoplastic (PVC) (solid aluminium or shaped copper conductors)	Armoured or non-armoured	Any	8
Mineral	Copper sheath with or without covering	Any	6‡

* For flat cables the diameter refers to the major axis.
† The value in brackets relates to single-core circular conductors of stranded construction installed in conduit, ducting or trunking.
‡ Mineral insulated cables may be bent to a radius not less than three times the cable diameter over the copper sheath, provided that the bend is not reworked, i.e. straightened and re-bent.

Appendix E
Cable capacities of conduit and trunking

A number of variable factors affect any attempt to arrive at a standard method of assessing the capacity of conduit or trunking.

Some of these are:

- reasonable care (of drawing-in)
- acceptable use of the space available
- tolerance in cable sizes
- tolerance in conduit and trunking.

The following tables can only give guidance on the maximum number of cables which should be drawn in. The sizes should ensure an easy pull with low risk of damage to the cables.

Only the ease of drawing-in is taken into account. The electrical effects of grouping are not. As the number of circuits increases the installed current-carrying capacity of the cable decreases. Cable sizes have to be increased with consequent increase in cost of cable and conduit.

It may sometimes be more attractive economically to divide the circuits concerned between two or more enclosures.

If thermosetting cables are installed in the same conduit or trunking as thermoplastic (PVC) insulated cables, the conductor operating temperature of any of the cables must not exceed that for thermoplastic (PVC), i.e. thermosetting cables must be rated as thermoplastic (PVC).

The following three cases are dealt with. Single-core thermoplastic (PVC) insulated cables in:

  i   straight runs of conduit not exceeding 3 m in length (Tables E1 and E2)
 ii   straight runs of conduit exceeding 3 m in length, or in runs of any length incorporating bends or sets (Tables E3 and E4)
iii   trunking (Tables E5 and E6).

For cables and/or conduits not covered by this appendix, advice on the number of cables that can be drawn in should be obtained from the manufacturer.

E Appendix

i Single-core thermoplastic (PVC) insulated cables in straight runs of conduit not exceeding 3 m in length

For each cable it is intended to use, obtain the appropriate factor from Table E1.

Add the cable factors together and compare the total with the conduit factors given in Table E2.

The minimum conduit size is that having a factor equal to or greater than the sum of the cable factors.

▼ **Table E1** Cable factors for use in conduit in short straight runs

Type of conductor	Conductor cross-sectional area (mm²)	Cable factor
Solid	1	22
	1.5	27
	2.5	39
Stranded	1.5	31
	2.5	43
	4	58
	6	88
	10	146
	16	202
	25	385

▼ **Table E2** Conduit factors for use in short straight runs

Conduit diameter (mm)	Conduit factor
16	290
20	460
25	800
32	1400
38	1900
50	3500
63	5600

Appendix E

ii Single-core thermoplastic (PVC) insulated cables in straight runs of conduit exceeding 3 m in length, or in runs of any length incorporating bends or sets

For each cable it is intended to use, obtain the appropriate factor from Table E3.

Add the cable factors together and compare the total with the conduit factors given in Table E4, taking into account the length of run it is intended to use and the number of bends and sets in that run.

The minimum conduit size is that size having a factor equal to or greater than the sum of the cable factors. For the larger sizes of conduit, multiplication factors are given relating them to 32 mm diameter conduit.

▼ **Table E3** Cable factors for use in conduit in long straight runs over 3 m, or runs of any length incorporating bends

Type of conductor	Conductor cross-sectional area (mm²)	Cable factor
Solid or Stranded	1	16
	1.5	22
	2.5	30
	4	43
	6	58
	10	105
	16	145
	25	217

The inner radius of a conduit bend should be not less than 2.5 times the outside diameter of the conduit.

E Appendix

▶ **Table E4** Conduit factors for runs incorporating bends and long straight runs

Length of run (m)	Straight				One Bend				Two Bends				Three Bends				Four Bends			
	16	20	25	32	16	20	25	32	16	20	25	32	16	20	25	32	16	20	25	32
1	Covered by Tables E1 and E2				188	303	543	947	177	286	514	900	158	256	463	818	130	213	388	692
1.5					182	294	528	923	167	270	487	857	143	233	422	750	111	182	333	600
2					177	286	514	900	158	256	463	818	130	213	388	692	97	159	292	529
2.5					171	278	500	878	150	244	442	783	120	196	358	643	86	141	260	474
3					167	270	487	857	143	233	422	750	111	182	333	600				
3.5	179	290	521	911	162	263	475	837	136	222	404	720	103	169	311	563				
4	177	286	514	900	158	256	463	818	130	213	388	692	97	159	292	529				
4.5	174	282	507	889	154	250	452	800	125	204	373	667	91	149	275	500				
5	171	278	500	878	150	244	442	783	120	196	358	643	86	141	260	474				
6	167	270	487	857	143	233	422	750	111	182	333	600								
7	162	263	475	837	136	222	404	720	103	169	311	563								
8	158	256	463	818	130	213	388	692	97	159	292	529								
9	154	250	452	800	125	204	373	667	91	149	275	500								
10	150	244	442	783	120	196	358	643	86	141	260	474								

Conduit diameter (mm)

Additional factors:
▲ For 38 mm diameter use 1.4 x (32 mm factor)
▲ For 50 mm diameter use 2.6 x (32 mm factor)
▲ For 63 mm diameter use 4.2 x (32 mm factor)

Appendix E

iii Single-core thermoplastic (PVC) insulated cables in trunking

For each cable it is intended to use, obtain the appropriate factor from Table E5.

Add the cable factors together and compare the total with the factors for trunking given in Table E6.

The minimum size of trunking is that size having a factor equal to or greater than the sum of the cable factors.

▼ **Table E5** Cable factors for trunking

Type of conductor	Conductor cross-sectional area (mm^2)	PVC BS 6004 Cable factor	Thermosetting BS 7211 Cable factor
Solid	1.5	8.0	8.6
	2.5	11.9	11.9
Stranded	1.5	8.6	9.6
	2.5	12.6	13.9
	4	16.6	18.1
	6	21.2	22.9
	10	35.3	36.3
	16	47.8	50.3
	25	73.9	75.4

Notes:
1 These factors are for metal trunking and may be optimistic for plastic trunking, where the cross-sectional area available may be significantly reduced from the nominal by the thickness of the wall material.
2 The provision of spare space is advisable; however, any circuits added at a later date must take into account grouping, Regulation 523.5.

Appendix

▼ **Table E6** Factors for trunking

Dimensions of trunking (mm x mm)	Factor	Dimensions of trunking (mm x mm)	Factor
50 x 38	767	200 x 100	8572
50 x 50	1037	200 x 150	13001
75 x 25	738	200 x 200	17429
75 x 38	1146	225 x 38	3474
75 x 50	1555	225 x 50	4671
75 x 75	2371	225 x 75	7167
100 x 25	993	225 x 100	9662
100 x 38	1542	225 x 150	14652
100 x 50	2091	225 x 200	19643
100 x 75	3189	225 x 225	22138
100 x 100	4252	300 x 38	4648
150 x 38	2999	300 x 50	6251
150 x 50	3091	300 x 75	9590
150 x 75	4743	300 x 100	12929
150 x 100	6394	300 x 150	19607
150 x 150	9697	300 x 200	26285
200 x 38	3082	300 x 225	29624
200 x 50	4145	300 x 300	39428
200 x 75	6359		

Note: Space factor is 45% with trunking thickness taken into account.

Other sizes and types of cable or trunking

For sizes and types of cable or trunking other than those given in Tables E5 and E6, the number of cables installed should be such that the resulting space factor does not exceed 45% of the net internal cross-sectional area.

Space factor is the ratio (expressed as a percentage) of the sum of the overall cross-sectional areas of cables (including insulation and any sheath) to the internal cross-sectional area of the trunking or other cable enclosure in which they are installed. The effective overall cross-sectional area of a non-circular cable is taken as that of a circle of diameter equal to the major axis of the cable.

Care should be taken to use trunking bends etc which do not impose bending radii on cables less than those required by Table D5.

Appendix F

Current-carrying capacities and voltage drop for copper conductors

Current-carrying capacity

523
435.1

In this simplified approach the assumption is made that the overcurrent protective device provides both fault current and overload current protection.

For cables buried in the ground, refer to BS 7671:2008(2011), Appendix 4.

Procedure

Appx 4, 3
433.1.1

1. The design current (I_b) of the circuit must first be established.
2. The overcurrent device rating (I_n) is then selected so that In is greater than or equal to I_b

$$I_n \geq I_b$$

The tabulated current-carrying capacity of the selected cable (I_t) is then given by:

$$I_t \geq \frac{I_n}{C_a \, C_g \, C_i \, C_f}$$

for simultaneously occurring factors.

C is a rating factor to be applied where the installation conditions differ from those for which values of current-carrying capacity are tabulated in this appendix. The various rating factors are identified as follows:

C_a for ambient temperature, see Table F1
C_g for grouping, see Table F3
C_i for thermal insulation, see Table F2 (Note: For cables installed in thermal insulation as described in Tables F4(i), F5(i) and F6, $C_i = 1$)

F Appendix

C_f for the type of protective device, i.e.:
- where the protective device is a semi-enclosed fuse to BS 3036, $C_f = 0.725$
- for all other devices $C_f = 1$.

Voltage drop

To calculate the voltage drop in volts the tabulated value of voltage drop (mV/A/m) has to be multiplied by the design current of the circuit (I_b), the length of run in metres (L), and divided by 1000 (to convert to volts):

$$\text{voltage drop} = \frac{(mV/A/m) \times I_b \times L}{1000}$$

The requirements of BS 7671 are deemed to be satisfied if the voltage drop between the origin of the installation and a lighting point does not exceed 3 per cent of the nominal voltage (6.9 V) and for other current-using equipment or socket-outlets does not exceed 5 per cent (11.5 V single-phase).

▼ **Table F1** Rating factors (C_a) for ambient air temperatures other than 30 °C to be applied to the current-carrying capacities for cables in free air

Ambient temperature (°C)	Insulation			
	70 °C thermoplastic	90 °C thermosetting	Mineral	
			Thermoplastic covered or bare and exposed to touch 70 °C	Bare and not exposed to touch 105 °C
25	1.03	1.02	1.07	1.04
30	1.00	1.00	1.00	1.00
35	0.94	0.96	0.93	0.96
40	0.87	0.91	0.85	0.92

Appendix F

523.9 Thermal insulation

Where a cable is to be run in a space to which thermal insulation is likely to be applied, the cable should, wherever practicable, be fixed in a position such that it will not be covered by the thermal insulation. Where fixing in such a position is impracticable, the cross-sectional area of the cable must be increased appropriately.

For a cable installed in thermal insulation as described in Tables F4(i), F5(i) and F6 no correction is required.

Note: Reference methods 100, 101 and 102 require the cable to be in contact with the plasterboard or the joists, see Tables 7.1(ii) and 7.1(iii) in Section 7.

For a single cable likely to be totally surrounded by thermally insulating material over a length of more than 0.5 m, the current-carrying capacity should be taken, in the absence of more precise information, as 0.5 times the current-carrying capacity for that cable clipped direct to a surface and open (reference method C).

Where a cable is totally surrounded by thermal insulation for less than 0.5 m the current-carrying capacity of the cable should be reduced appropriately depending on the size of cable, length in insulation and thermal properties of the insulation. The derating factors in Table F2 are appropriate to conductor sizes up to 10 mm² in thermal insulation having a thermal conductivity (λ) greater than 0.04 Wm⁻¹K⁻¹.

Table 52.2

▼ **Table F2** Cable surrounded by thermal insulation

Length in insulation (mm)	Derating factor (C_i)
50	0.88
100	0.78
200	0.63
400	0.51
≥ 500	0.50

Appendix

▶ **Table F3** Rating factors (C_g) for one circuit or one multicore cable or for a group of circuits, or a group of multicore cables (to be used with the current-carrying capacities of Tables F4(i), F5(i) and F6)

Table 4C1

Arrangement (cables touching)	Number of circuits or multicore cables										Applicable reference method for current-carrying capacities
	1	2	3	4	5	6	7	8	9	12	
Bunched in air, on a surface, embedded or enclosed	1.0	0.80	0.70	0.65	0.60	0.57	0.54	0.52	0.50	0.45	A to F
Single layer on wall or floor	1.0	0.85	0.79	0.75	0.73	0.72	0.72	0.71	0.70	0.70	C
Single layer multicore on a perforated horizontal or vertical cable tray system	1.0	0.88	0.82	0.77	0.75	0.73	0.73	0.72	0.72	0.72	E
Single layer multicore on a cable ladder system or cleats, etc.	1.0	0.87	0.82	0.80	0.80	0.79	0.79	0.78	0.78	0.78	E

Appendix F

Notes to Table F3:

1 These factors are applicable to uniform groups of cables, equally loaded.

2 Where horizontal clearances between adjacent cables exceed twice their overall diameter, no rating factor need be applied.

3 The same factors are applied to:
   ▲ groups of two or three single-core cables
   ▲ multicore cables.

4 If a group consists of both two- and three-core cables, the total number of cables is taken as the number of circuits, and the corresponding factor is applied to the tables for two loaded conductors for the two-core cables, and to the tables for three loaded conductors for the three-core cables.

5 If a group consists of n single-core cables it may either be considered as $n/2$ circuits of two loaded conductors (for single-phase circuits) or $n/3$ circuits of three loaded conductors (for three-phase circuits).

6 The rating factors given have been averaged over the range of conductor sizes and types of installation included in Tables 4D1A to 4J4A of BS 7671 (this includes F4(i), F5(i) and F6 of this guide) and the overall accuracy of tabulated values is within 5%.

7 For some installations and for other methods not provided for in the above table, it may be appropriate to use factors calculated for specific cases, see for example Tables 4C4 and 4C5 of BS 7671.

8 Where cables having differing conductor operating temperature are grouped together, the current rating is to be based upon the lowest operating temperature of any cable in the group.

523.5   9 If, due to known operating conditions, a cable is expected to carry not more than 30% of its grouped rating, it may be ignored for the purpose of obtaining the rating factor for the rest of the group. For example, a group of N loaded cables would normally require a group rating factor of Cg applied to the tabulated It. However, if M cables in the group carry loads which are not greater than 0.3 CgIt amperes the other cables can be sized by using the group rating factor corresponding to (N minus M) cables.

F | Appendix

▶ **Table F4(i)** Single-core 70 °C thermoplastic (PVC) or thermosetting (note 1) insulated cables, non-armoured, with or without sheath (copper conductors)
Table 4D1A

Ambient temperature: 30 °C
Conductor operating temperature: 70 °C

Current-carrying capacity (amperes):

Conductor cross-sectional area	Reference method A (enclosed in conduit in thermally insulating wall, etc.)		Reference method B (enclosed in conduit on a wall or in trunking, etc.)		Reference method C (clipped direct)			Reference method F (in free air or on a perforated cable tray horizontal or vertical)					
										Touching		Spaced by one cable diameter	
	2 cables, single-phase a.c. or d.c.	3 or 4 cables, three-phase a.c.	2 cables, single-phase a.c. or d.c.	3 or 4 cables, three-phase a.c.	2 cables, single-phase a.c. or d.c flat and touching	3 or 4 cables, three-phase a.c. flat and touching or trefoil		2 cables, single-phase a.c. or d.c. flat	3 cables, three-phase a.c. flat	3 cables three-phase a.c. trefoil	2 cables single-phase a.c. or d.c. or 3 cables three-phase a.c. flat	
												horizontal	vertical
mm²	1 A	2 A	3 A	4 A	5 A	6 A	7 A	8 A	9 A	10 A	11 A	12 A
1	11	10.5	13.5	12	15.5	14						
1.5	14.5	13.5	17.5	15.5	20	18						
2.5	20	18	24	21	27	25						
4	26	24	32	28	37	33						
6	34	31	41	36	47	43						
10	46	42	57	50	65	59						
16	61	56	76	68	87	79						
25	80	73	101	89	114	104	131	114	110	146	130	

Appendix F

▼ **Table F4(i)** continued

Conductor cross-sectional area	Reference method A (enclosed in conduit in thermally insulating wall, etc.)		Reference method B (enclosed in conduit on a wall or in trunking, etc.)		Reference method C (clipped direct)			Reference method F (in free air or on a perforated cable tray horizontal or vertical)				
	2 cables, single phase a.c. or d.c.	3 or 4 cables, three phase a.c.	2 cables, single phase a.c. or d.c.	3 or 4 cables, three phase a.c.	2 cables, single-phase a.c. or d.c. flat and touching	3 or 4 cables, three-phase a.c. flat and touching or trefoil		Touching		Spaced by one cable diameter		
							2 cables, single phase a.c. or d.c. flat	3 cables, three-phase a.c. flat	3 cables, three-phase a.c. trefoil	2 cables single-phase a.c. or d.c. or 3 cables three-phase a.c. flat		
										horizontal	horizontal	vertical
1	2	3	4	5	6	7	8	9	10	11	11	12
mm²	A	A	A	A	A	A	A	A	A	A	A	A
35	99	89	125	110	141	129	162	143	137	181	181	162
50	119	108	151	134	182	167	196	174	167	219	219	197
70	151	136	192	171	234	214	251	225	216	281	281	254
95	182	164	232	207	284	261	304	275	264	341	341	311

Notes to Table F4(i):

1. The ratings for cables with thermosetting insulation are applicable for cables connected to equipment or accessories designed to operate with cables which run at a temperature not exceeding 70 °C. Where conductor operating temperatures up to 90 °C are acceptable the current rating is increased – see Table 4E1A of BS 7671. Where the conductor is to be protected by a semi-enclosed fuse to BS 3036, see the introduction to this appendix.
2. The current-carrying capacities in columns 2 to 5 are also applicable to flexible cables to BS 6004 Table 1(c) and to 90 °C heat-resisting PVC cables to BS 6231 Tables 8 and 9 where the cables are used in fixed installations.

F | Appendix

▼ **Table F4(ii)** Voltage drop (per ampere per metre) at a conductor operating temperature of 70 °C
Table 4D1B

Conductor cross-sectional area	2 cables d.c.	2 cables, single-phase a.c.			Reference methods A & B (enclosed in conduit or trunking)	3 or 4 cables, three-phase a.c.		
		Reference methods A & B (enclosed in conduit or trunking)	Reference methods C & F (clipped direct on tray or in free air) touching	Reference methods C & F (clipped direct on tray or in free air) spaced		Reference methods C & F (clipped direct, on tray or in free air) Touching. Trefoil	Reference methods C & F (clipped direct, on tray or in free air) Touching. Flat	Reference methods C & F (clipped direct, on tray or in free air) Spaced*, Flat
mm²	mV/A/m	mV/A/m	mV/A/m	mV/A/m	mV/A/m	mV/A/m	mV/A/m	mV/A/m
1	44	44	44	44	38	38	38	38
1.5	29	29	29	29	25	25	25	25
2.5	18	18	18	18	15	15	15	15
4	11	11	11	11	9.5	9.5	9.5	9.5
6	7.3	7.3	7.3	7.3	6.4	6.4	6.4	6.4
10	4.4	4.4	4.4	4.4	3.8	3.8	3.8	3.8
16	2.8	2.8	2.8	2.8	2.4	2.4	2.4	2.4
	z†	z†	z†	z†	z†	z†	z†	z†
25	1.75	1.80	1.75	1.80	1.55	1.50	1.55	1.55
35	1.25	1.30	1.25	1.30	1.10	1.10	1.10	1.15
50	0.93	1.00	0.95	0.97	0.85	0.82	0.84	0.86
70	0.63	0.72	0.66	0.69	0.61	0.57	0.60	0.63
95	0.46	0.56	0.50	0.54	0.48	0.43	0.47	0.51

* Spacings larger than one cable diameter will result in larger voltage drop.
† The impedance values in Table F4(ii) consist of both the resistive and reactive elements of voltage drop, usually provided separately for 25 mm² and above conductor sizes.

For more information, see Appendix 4 of BS 7671.

Appendix F

▼ **Table F5(i)** Multicore cables having thermoplastic (PVC) or thermosetting insulation (note 1), non-armoured (copper conductors)

Table 4D2A

Ambient temperature: 30 °C
Conductor operating temperature: 70 °C

Current-carrying capacity (amperes):

Conductor cross-sectional area	Reference method A (enclosed in conduit in a thermally insulating wall, etc.)		Reference method B (enclosed in conduit on a wall or in trunking, etc.)		Reference method C (clipped direct)		Reference method E (in free air or on a perforated cable tray, etc. horizontal or vertical)	
	1 two-core cable*, single-phase a.c. or d.c.	1 three-core cable* or 1 four-core cable, three-phase a.c.	1 two-core cable*, single-phase a.c. or d.c.	1 three-core cable* or 1 four-core cable, three-phase a.c.	1 two-core cable*, single-phase a.c. or d.c.	1 three-core cable* or 1 four-core cable, three-phase a.c.	1 two-core cable*, single-phase a.c. or d.c.	1 three-core cable* or 1 four-core cable, three-phase a.c.
	2	3	4	5	6	7	8	9
mm²	A	A	A	A	A	A	A	A
1	11	10	13	11.5	15	13.5	17	14.5
1.5	14	13	16.5	15	19.5	17.5	22	18.5
2.5	18.5	17.5	23	20	27	24	30	25
4	25	23	30	27	36	32	40	34
6	32	29	38	34	46	41	51	43
10	43	39	52	46	63	57	70	60
16	57	52	69	62	85	76	94	80
25	75	68	90	80	112	96	119	101
35	92	83	111	99	138	119	148	126

Appendix F

▶ **Table F5(i)** continued

Conductor cross-sectional area	Reference method A (enclosed in conduit in a thermally insulating wall, etc.)		Reference method B (enclosed in conduit on a wall or in trunking, etc.)		Reference method C (clipped direct)		Reference method E (in free air or on a perforated cable tray, etc. horizontal or vertical)	
	1 two-core cable*, single-phase a.c. or d.c.	1 three-core cable* or 1 four-core cable, three-phase a.c.	1 two-core cable*, single-phase a.c. or d.c.	1 three-core cable* or 1 four-core cable, three-phase a.c.	1 two-core cable*, single-phase a.c. or d.c.	1 three-core cable* or 1 four-core cable, three-phase a.c.	1 two-core cable*, single-phase a.c. or d.c.	1 three-core cable* or 1 four-core cable, three-phase a.c.
1	2	3	4	5	6	7	8	9
mm²	A	A	A	A	A	A	A	A
50	110	99	133	118	168	144	180	153
70	139	125	168	149	213	184	232	196
95	167	150	201	179	258	223	282	238

Notes to Table F5(i):

1. The ratings for cables with thermosetting insulation are applicable for cables connected to equipment or accessories designed to operate with cables which run at a temperature not exceeding 70 °C. Where conductor operating temperatures up to 90 °C are acceptable the current rating is increased – see Table 4E2A of BS 7671. Where the conductor is to be protected by a semi-enclosed fuse to BS 3036, see the introduction to this appendix.
2. With or without protective conductor. Circular conductors are assumed for sizes up to and including 16 mm². Values for larger sizes relate to shaped conductors and may safely be applied to circular conductors.

*

Appendix F

Table 4D2B

Table F5(ii) Voltage drop (per ampere per metre) at a conductor operating temperature of 70 °C

Conductor cross-sectional area	Two-core cable, d.c.	Two-core cable, single-phase a.c.	Three- or four-core cable, three-phase
1	2	3	4
mm²	mV/A/m	mV/A/m	mV/A/m
1	44	44	38
1.5	29	29	25
2.5	18	18	15
4	11	11	9.5
6	7.3	7.3	6.4
10	4.4	4.4	3.8
16	2.8	2.8	2.4
		z†	z†
25	1.75	1.75	1.50
35	1.25	1.25	1.10
50	0.93	0.94	0.81
70	0.63	0.65	0.57
95	0.46	0.50	0.43

† The impedance values in Table F5(ii) consist of both the resistive and reactive elements of voltage drop, usually provided separately for 25 mm² and above conductor sizes.
For more information, see Appendix 4 of BS 7671.

F | Appendix

▶ **Table F6** 70 °C thermoplastic (PVC) insulated and sheathed flat cable with protective conductor (copper conductors)
Table 4D5

Ambient temperature: 30 °C
Conductor operating temperature: 70 °C

Current-carrying capacity (amperes) and voltage drop (per ampere per metre):

Conductor cross-sectional area	Reference method 100* (above a plasterboard ceiling covered by thermal insulation not exceeding 100 mm in thickness)	Reference method 101* (above a plasterboard ceiling covered by thermal insulation exceeding 100 mm in thickness)	Reference method 102* (in a stud wall with thermal insulation with cable touching the inner wall surface)	Reference method 103 (in a stud wall with thermal insulation with cable not touching the inner wall surface)	Reference method C (clipped direct)	Reference method A (enclosed in conduit in an insulated wall)	Voltage drop
1	2	3	4	5	6	7	8
mm²	A	A	A	A	A	A	mV/A/m
1	13	10.5	13	8	16	11.5	44
1.5	16	13	16	10	20	14.5	29
2.5	21	17	21	13.5	27	20	18
4	27	22	27	17.5	37	26	11
6	34	27	35	23.5	47	32	7.3
10	45	36	47	32	64	44	4.4
16	57	46	63	42.5	85	57	2.8

Notes:

* Reference methods 100, 101 and 102 require the cable to be in contact with the plasterboard ceiling, wall or joist, see Tables 7.1(ii) and 7.1(iii) in Section 7.
1. Wherever practicable, a cable is to be fixed in a position such that it will not be covered with thermal insulation.
2. Regulation 523.9, BS 5803-5: Appendix C: Avoidance of overheating of electric cables, Building Regulations Approved Document B and Thermal insulation: avoiding risks, BR 262, BRE, 2001 refer.

Appendix G
Certification and reporting

The certificates and forms are used with the kind permission of BSI.

G1 Introduction
Fundamentally, two types of form are recognised by BS 7671, certificates and reports:
- ▶ certificates are issued for new installation work
- ▶ reports are issued for inspections of existing installations.

G2 Certification
Two types of certificate for new work are recognised by BS 7671:
- ▶ Electrical Installation Certificate
- ▶ Minor Electrical Installation Works Certificate.

G2.1 Electrical Installation Certificate
The Electrical Installation Certificate is intended to be issued where more significant installation work is undertaken; examples are:
- ▶ a complete installation for a new property
- ▶ rewire of an existing installation
- ▶ replacement of a consumer unit
- ▶ addition of a new circuit from the distribution board or consumer unit.

G2.2 Minor Electrical Installation Works Certificate
The Minor Electrical Installation Works Certificate is intended to be issued for an addition or alteration to an existing circuit; examples are:
- ▶ adding lights to a lighting circuit
- ▶ adding socket-outlets to a ring final circuit
- ▶ rerouting an existing circuit
- ▶ replacing an existing shower with a larger power rating of unit
- ▶ replacing circuit-breakers with RCBOs where there is a difference of overcurrent type, e.g. replacing Type C for Type B.

G Appendix

In each case, the *characteristics* of the circuit are likely to have been altered, whether it's the addition of extra load or changes to the original earth fault loop impedance.

G2.3 Accountability

Certificates call for those responsible for the electrical installation or construction work to certify that the requirements of the Regulations have been met. Under no circumstances should a third party issue a certificate for installation work they have not undertaken.

It is common with larger installations for the design to be carried out by one company, installation or construction by someone else and the inspection and testing to be undertaken by some other, e.g. a testing organisation working on behalf of the installer; this is quite acceptable but the company who carries out the installation must issue the Electrical Installation Certificate.

G3 Reporting

G3.1 Electrical Installation Condition Report

The Electrical Installation Condition Report (EICR) is intended to be issued when a periodic inspection of an electrical installation has been carried out. The EICR does not certify anything and, hence, must not be issued to certify new electrical installation work. The purpose of the EICR is to report on the condition of an existing electrical installation and, ultimately, present one of two outcomes:

- ▶ SATISFACTORY – the installation is deemed safe for continued use
- ▶ UNSATISFACTORY – one or more issues of safety have been identified.

Where an unsatisfactory result has been recorded, C1 and/or C2 observations will have been included identifying the reason(s) for the result. Once the report has been issued by the inspector, the onus is then placed on the client to act in response to the observations recorded.

G3.2 Observations

Observations to be recorded fall into three categories:

C1 – Danger present. Risk of injury. Immediate remedial action required
C2 – Potentially dangerous – urgent remedial action required
C3 – Improvement recommended.

Examples of C1

Where danger currently exists and an immediate issue of safety is apparent:

- ▶ uninsulated live conductors exposed on broken wiring accessory
- ▶ incorrect polarity at socket-outlets, e.g. live/cpc reversal
- ▶ item of metalwork that has become live due to a fault.

Examples of C2

Not immediately dangerous but a dangerous condition could occur due to a fault:

- ▶ main equipotential bonding not installed to extraneous-conductive-parts
- ▶ RCD (30 mA for additional protection) fails to operate in the required time

Appendix G

- double-pole fusing (line and neutral)
- no connection to means of earthing at origin
- no cpc for lighting circuit having Class I fittings/accessories with exposed-conductive-parts.

Examples of C3
Installations complying with older versions of BS 7671:
- no 30 mA RCD for additional protection for socket-outlets used by unskilled/uninstructed persons
- earth leakage circuit-breaker installed at origin of TT installation
- no cpc for lighting circuit where only Class II fittings/accessories are installed.

G3.3 Dangerous situations

Where the inspector discovers an extremely dangerous situation, e.g. persons or livestock are at immediate risk of electric shock or an imminent fire hazard is evident, urgent action is advised to remove the danger. As the expert, the inspector has been employed to identify electrical problems and, therefore, should make safe such dangerous issues while on the premises.

The inspector is advised to exercise judgement to secure the area and inform the client immediately, followed up in writing. Once permission has been obtained, the danger should be removed.

G3.4 Remedial work

Often the client will ask how much time they have before any necessary remedial work should be carried out once alerted of the unsatisfactory result of the inspection. There is no standard answer that can be given as all installations and situations are different from each other. It is worth informing the client, however, that the installation has been given an unsatisfactory result as there are issues of electrical safety and a duty of care exists in law to ensure that employees or members of the public are not placed in a position of unacceptable risk.

When remedial work has been completed in response to the findings of a periodic inspection, the work will need to be certified as described in G2.

G4 Introduction to Model Forms from BS 7671:2008(2011)

For convenience, the forms are numbered as below:

Form 1	Electrical Installation Certificate (single-signature)
Form 2	Electrical Installation Certificate
Form 3	Schedule of Inspections
Form 4	Generic Schedule of Test Results
Form 5	Minor Electrical Installation Works Certificate
Form 6	Electrical Installation Condition Report
Form 7	Condition Report Inspection Schedule

G | Appendix

Appx 6　The introduction to Appendix 6 'Model forms for certification and reporting' of BS 7671:2008(2011) is reproduced below.

(i) The Electrical Installation Certificate required by Part 6 should be made out and signed or otherwise authenticated by a competent person or persons in respect of the design, construction, inspection and testing of the work.

(ii) The Minor Works Certificate required by Part 6 should be made out and signed or otherwise authenticated by a competent person in respect of the design, construction, inspection and testing of the minor work.

(iii) The Electrical Installation Condition Report required by Part 6 should be made out and signed or otherwise authenticated by a competent person in respect of the inspection and testing of an installation.

(iv) Competent persons will, as appropriate to their function under (i) (ii) and (iii) above, have a sound knowledge and experience relevant to the nature of the work undertaken and to the technical standards set down in these Regulations, be fully versed in the inspection and testing procedures contained in these Regulations and employ adequate testing equipment.

(v) Electrical Installation Certificates will indicate the responsibility for design, construction, inspection and testing, whether in relation to new work or further work on an existing installation.

Where design, construction, inspection and testing are the responsibility of one person a Certificate with a single-signature declaration in the form shown below may replace the multiple signatures section of the model form.

FOR DESIGN, CONSTRUCTION, INSPECTION & TESTING

I being the person responsible for the Design, Construction, Inspection & Testing of the electrical installation (as indicated by my signature below), particulars of which are described above, having exercised reasonable skill and care when carrying out the Design, Construction, Inspection & Testing, hereby CERTIFY that the said work for which I have been responsible is to the best of my knowledge and belief in accordance with BS 7671:2008, amended to ....(date) except for the departures, if any, detailed as follows.

(vi) A Minor Works Certificate will indicate the responsibility for design, construction, inspection and testing of the work described on the certificate.

(vii) An Electrical Installation Condition Report will indicate the responsibility for the inspection and testing of an existing installation within the extent and limitations specified on the report.

(viii) Schedules of inspection and schedules of test results as required by Part 6 should be issued with the associated Electrical Installation Certificate or Electrical Installation Condition Report.

(ix) When making out and signing a form on behalf of a company or other business entity, individuals should state for whom they are acting.

(x) Additional forms may be required as clarification, if needed by ordinary persons, or in expansion, for larger or more complex installations.

(xi) The IET Guidance Note 3 provides further information on inspection and testing and for periodic inspection, testing and reporting.

Appendix G

G4.1 Electrical Installation Certificate

Figures G4.1(i)–(iv) show a typical completed Electrical Installation Certificate comprising Forms 1, 3 and 4. It is assumed that the diagrams and documentation required by Regulation 514.9 are available. The installation is for a music shop, which has SELV lighting, wiring in close proximity to gas pipes, broadband and data cables, and has fire sealed trunking through to a store room. Regarding Form 4, the continuity test has been carried out using $(R_1 + R_2)$ and hence R_2 testing is Not Applicable; also, since RCDs 1 and 2 each protect three circuits "ditto" marks have been used on the form. Different test instruments will show different displays indicating "out of range", e.g. +299 or >199.

G | Appendix

▼ **Figure G4.1(i)** Electrical Installation Certificate – page 1

Form 1
Form No: S.V.T.-1.../1
ELECTRICAL INSTALLATION CERTIFICATE
(REQUIREMENTS FOR ELECTRICAL INSTALLATIONS - BS 7671 [IET WIRING REGULATIONS])

DETAILS OF THE CLIENT Buzz Music Store
22 Johnston Street
Oldtown .... Post Code: AC30 1DC

INSTALLATION ADDRESS
Buzz Music Store
22 Johnston Street
Oldtown .... Post Code: AC30 1DC

DESCRIPTION AND EXTENT OF THE INSTALLATION Tick boxes as appropriate

New installation ☑

Description of installation:
Rewire of small commercial premises - music shop

Addition to an existing installation ☐

Extent of installation covered by this Certificate:
Complete installation

Alteration to an existing installation ☐

(Use continuation sheet if necessary) see continuation sheet No: N/A

FOR DESIGN, CONSTRUCTION, INSPECTION & TESTING
I being the person responsible for the design, construction, inspection & testing of the electrical installation (as indicated by my signature below), particulars of which are described above, having exercised reasonable skill and care when carrying out the design, construction, inspection & testing hereby CERTIFY that the said work for which I have been responsible is to the best of my knowledge and belief in accordance with BS 7671:2008, amended to 2011 (date) except for the departures, if any, detailed as follows:

Details of departures from BS 7671 (Regulations 120.3 and 133.5):
None

The extent of liability of the signatory is limited to the work described above as the subject of this Certificate.
Signature: *Clive Jenkin* Date: 4-Jan-2012 Name (IN BLOCK LETTERS): CLIVE JENKIN
Company: PCJ Electrical
Address: 22 Whinlatter Close,
Oldtown Postcode: AC30 8CD Tel No: 01234 567890

NEXT INSPECTION
I recommend that this installation is further inspected and tested after an interval of not more than 5 years/~~months~~.

SUPPLY CHARACTERISTICS AND EARTHING ARRANGEMENTS Tick boxes and enter details, as appropriate

Earthing arrangements	Number and Type of Live Conductors	Nature of Supply Parameters	Supply Protective Device Characteristics
TN-C ☐ TN-S ☐ TN-C-S ☑ TT ☐ IT ☐	a.c. ☑ d.c. ☐ 1-phase, 2-wire ☑ 2-wire ☐ 1-phase, 3-wire ☐ 3-wire ☐ 2-phase, 3-wire ☐ other ☐ 3-phase, 3-wire ☐ 3-phase, 4-wire ☐	Nominal voltage, $U/U_0^{(1)}$ .... 230 V Nominal frequency, $f^{(1)}$ .... 50 Hz Prospective fault current, $I_{pf}^{(2)}$ 1.09 kA External loop impedance, $Z_e^{(2)}$ 0.21 Ω *(Note: (1) by enquiry, (2) by enquiry or by measurement)*	Type: BS 1361 Fuse Rated current .... 100 A
Other sources ☐ of supply (to be detailed on attached schedules)	Confirmation of supply polarity ☑		

Page 1 of 4

Appendix G

▼ **Figure G4.1(ii)** Electrical Installation Certificate – page 2

Form 1   Form No: SVT -1..../1

PARTICULARS OF INSTALLATION REFERRED TO IN THE CERTIFICATE Tick boxes and enter details, as appropriate

Means of Earthing	Maximum Demand		
Distributor's facility ☑	Maximum demand (load) .... 60 kVA / Amps *Delete as appropriate*		
	Details of Installation Earth Electrode (*where applicable*)		
Installation earth electrode ☐	Type (e.g. rod(s), tape etc) N/A	Location N/A	Electrode resistance to Earth N/A Ω

Main Protective Conductors

Earthing conductor: material Copper   csa ....16....mm²   Continuity and connection verified ☑

Main protective bonding conductors   material Copper   csa ....10....mm²   Continuity and connection verified ☑

To incoming water and/or gas service ☑   To other elements: N/A..

Main Switch or Circuit-breaker

BS, Type and No. of poles BS EN 60947-3 (2-pole)   Current rating ....100..A   Voltage rating ....230....V

Location Storeroom (understairs)..........................   Fuse rating or setting.........N/A..A

Rated residual operating current $I_{\Delta n}$ = ...N/A.. mA, and operating time of ...N/A.. ms (at $I_{\Delta n}$) *(applicable only where an RCD is suitable and is used as a main circuit-breaker)*

COMMENTS ON EXISTING INSTALLATION (in the case of an addition or alteration see Section 633):
N/A...
....
....

SCHEDULES
The attached Schedules are part of this document and this Certificate is valid only when they are attached to it.
........1...... Schedules of Inspections and .....1...... Schedules of Test Results are attached.
(Enter quantities of schedules attached).

ELECTRICAL INSTALLATION CERTIFICATE
GUIDANCE FOR RECIPIENTS

This safety Certificate has been issued to confirm that the electrical installation work to which it relates has been designed, constructed, inspected and tested in accordance with British Standard 7671 (the IET Wiring Regulations).

You should have received an "original" Certificate and the contractor should have retained a duplicate. If you were the person ordering the work, but not the owner of the installation, you should pass this Certificate, or a full copy of it including the schedules, immediately to the owner.

The "original" Certificate should be retained in a safe place and be shown to any person inspecting or undertaking further work on the electrical installation in the future. If you later vacate the property, this Certificate will demonstrate to the new owner that the electrical installation complied with the requirements of British Standard 7671 at the time the Certificate was issued. The Construction (Design and Management) Regulations require that, for a project covered by those Regulations, a copy of this Certificate, together with schedules, is included in the project health and safety documentation.

For safety reasons, the electrical installation will need to be inspected at appropriate intervals by a competent person. The maximum time interval recommended before the next inspection is stated on Page 1 under "NEXT INSPECTION".

This Certificate is intended to be issued only for a new electrical installation or for new work associated with an addition or alteration to an existing installation. It should not have been issued for the inspection of an existing electrical installation. An "Electrical Installation Condition Report" should be issued for such an inspection.

G | Appendix

▼ **Figure G4.1(iii)** Schedule of Inspections - Electrical Installation Certificate – page 3

Form 3   Form No: SVT-1.../3

SCHEDULE OF INSPECTIONS (for new installation work only)

Methods of protection against electric shock

Both basic and fault protection:
- [✓] (i) SELV (note 1)
- [N/A] (ii) PELV
- [N/A] (iii) Double insulation
- [N/A] (iv) Reinforced insulation

Basic protection: (note 2)
- [✓] (i) Insulation of live parts
- [✓] (ii) Barriers or enclosures
- [N/A] (iii) Obstacles (note 3)
- [N/A] (iv) Placing out of reach (note 4)

Fault protection:

(i) Automatic disconnection of supply:
- [✓] Presence of earthing conductor
- [✓] Presence of circuit protective conductors
- [✓] Presence of protective bonding conductors
- [N/A] Presence of supplementary bonding conductors
- [N/A] Presence of earthing arrangements for combined protective and functional purposes
- [N/A] Presence of adequate arrangements for other source/s, where applicable
- [N/A] FELV
- [✓] Choice and setting of protective and monitoring devices (for fault and/or overcurrent protection)

(ii) **Non-conducting location:** (note 5)
- [N/A] Absence of protective conductors

(iii) **Earth-free local equipotential bonding:** (note 6)
- [N/A] Presence of earth-free local equipotential bonding

(iv) **Electrical separation:** (note 7)
- [N/A] Provided for **one item** of current-using equipment
- [N/A] Provided for **more than one item** of current-using equipment

Additional protection:
- [✓] Presence of residual current devices(s)
- [N/A] Presence of supplementary bonding conductors

Prevention of mutual detrimental influence
- [✓] (a) Proximity to non-electrical services and other influences
- [✓] (b) Segregation of Band I and Band II circuits or use of Band II insulation
- [N/A] (c) Segregation of safety circuits

Identification
- [✓] (a) Presence of diagrams, instructions, circuit charts and similar information
- [✓] (b) Presence of danger notices and other warning notices
- [✓] (c) Labelling of protective devices, switches and terminals
- [✓] (d) Identification of conductors

Cables and conductors
- [✓] Selection of conductors for current-carrying capacity and voltage drop
- [✓] Erection methods
- [✓] Routing of cables in prescribed zones
- [N/A] Cables incorporating earthed armour or sheath, or run within an earthed wiring system, or otherwise adequately protected against nails, screws and the like
- [✓] Additional protection provided by 30 mA RCD for cables concealed in walls (where required in premises not under the supervision of a skilled or instructed person)
- [✓] Connection of conductors
- [✓] Presence of fire barriers, suitable seals and protection against thermal effects

General
- [✓] Presence and correct location of appropriate devices for isolation and switching
- [✓] Adequacy of access to switchgear and other equipment
- [N/A] Particular protective measures for special installations and locations
- [✓] Connection of single-pole devices for protection or switching in line conductors only
- [✓] Correct connection of accessories and equipment
- [N/A] Presence of undervoltage protective devices
- [✓] Selection of equipment and protective measures appropriate to external influences
- [✓] Selection of appropriate functional switching devices

Inspected by _Clive Jenkins_   Date _4-Jan-2012_

NOTES:
✓ to indicate an inspection has been carried out and the result is satisfactory
N/A to indicate that the inspection is not applicable to a particular item
An entry must be made in every box.

1. SELV An extra-low voltage system which is electrically separated from Earth and from other systems. The particular requirements of the Regulations must be checked (see Section 414)
2. Method of basic protection - will include measurement of distances where appropriate
3. Obstacles - only adopted in special circumstances (see Regulations 417.1 and 417.2)
4. Placing out of reach - only adopted in special circumstances (see Regulations 417.1 and 417.3)
5. Non-conducting locations - not applicable in domestic premises and requiring special precautions (see Regulation 418.1)
6. Earth-free local equipotential bonding - not applicable in domestic premises, only used in special circumstances (see Regulation 418.2)
7. Electrical separation (see Section 413 and Regulation 418.3)

Page .3. of .4.

Appendix G

Figure G4.1(iv) Generic schedule of test results - Electrical Installation Certificate – page 4

Form 4

Form No: SYT.1.../4

GENERIC SCHEDULE OF TEST RESULTS

DB reference no: CU1
Location: Storeroom (understairs)
Zs at DB (Ω): 0.21
Ipf at DB (kA): 1.09
Correct supply polarity confirmed: ☑
Phase sequence confirmed (where appropriate): N/A

Details of circuits and/or installed equipment vulnerable to damage when testing: SELV lights installed on circuit #6

Tested by:
Name (Capitals): CLIVE JENKIN
Signature: Clive Jenkin
Date: 4 Jan 2011

Details of test instruments used (state serial and/or asset numbers)
- Continuity: Ser. No. 11638475
- Insulation resistance: Ser. No. 11638475
- Earth fault loop impedance: Ser. No. 11638475
- RCD: Ser. No. 11638475
- Earth electrode resistance: N/A

Circuit number	Circuit Description	\[Circuit details\] BS (EN)	type	rating (A)	breaking capacity (kA)	Reference Method	Live (mm²)	cpc (mm²)	Ring final circuit continuity (Ω) r₁ (line)	rₙ (neutral)	r₂ (cpc)	Continuity (Ω) (R₁+R₂) or R₂ *	R₂	Insulation Resistance (MΩ) Live-Live	Live-E	Polarity	Zs (Ω)	RCD I∆n (ms)	@ 5I∆n	Test button / functionality	Remarks (continue on a separate sheet if necessary)
1	Ring - sockets shop area	60898	B	32	6	C	2.5	1.5	0.6	0.6	0.97	0.38	N/A	+299	+299	✓	0.57	56	14.2	✓	RCD 1
2	Radial - Water heater	60898	B	16	6	C	2.5	1.5	N/A	N/A	N/A	0.2	N/A	+299	+299	✓	0.4	"	"	✓	RCD 1
3	Radial - Burglar alarm	60898	B	16	6	C	2.5	1.5	N/A	N/A	N/A	0.29	N/A	+299	+299	✓	0.49	"	"	✓	RCD 1
4	Radial - sockets sales counter and back room	60898	B	16	6	C	2.5	1.5	N/A	N/A	N/A	0.59	N/A	+299	+299	✓	0.78	48	19.9	✓	RCD 2
5	Radial - Fire alarm	60898	B	16	6	C	2.5	1.5	N/A	N/A	N/A	0.33	N/A	+299	+299	✓	0.53	"	"	✓	RCD 2
6	Lights - internal, external sign	60898	B	6	6	C	1.5	1.0	N/A	N/A	N/A	0.51	N/A	+299	+299	✓	0.71	"	"	✓	RCD 2
7	Spare																				
8	Spare																				

* Where there are no spurs connected to a ring final circuit this value is also the (R₁ + R₂) of the circuit.

NOTE: One schedule of test results will be issued for every consumer unit or distribution board

Page 4 of 4

G | Appendix

G4.2 Electrical Installation Certificate – Completion

Notes:

1. The Electrical Installation Certificate is to be used only for the initial certification of a new installation or for an addition or alteration to an existing installation where new circuits have been introduced.
   It is not to be used for a Periodic Inspection, for which an Electrical Inspection Condition Report form should be used.
   For an addition or alteration which does not extend to the introduction of new circuits, a Minor Electrical Installation Works Certificate may be used.
   The "original" Certificate is to be given to the person ordering the work (Regulation 632.1). A duplicate should be retained by the contractor.
2. This Certificate is only valid if accompanied by the Schedule of Inspections and the Schedule(s) of Test Results.
3. The signatures appended are those of the persons authorized by the companies executing the work of design, construction, inspection and testing respectively. A signatory authorized to certify more than one category of work should sign in each of the appropriate places.
4. The time interval recommended before the first periodic inspection must be inserted (see IET Guidance Note 3 for guidance).
5. The page numbers for each of the Schedules of Test Results should be indicated, together with the total number of sheets involved.
6. The maximum prospective value of fault current (I_{pf}) recorded should be the greater of either the prospective value of short-circuit current or the prospective value of earth fault current.
7. The proposed date for the next inspection should take into consideration the frequency and quality of maintenance that the installation can reasonably be expected to receive during its intended life, and the period should be agreed between the designer, installer and other relevant parties.

G4.3 Electrical Installation Certificate – Guidance for recipients

(to be appended to the Certificate)

This safety Certificate has been issued to confirm that the electrical installation work to which it relates has been designed, constructed, inspected and tested in accordance with British Standard 7671 (the IET Wiring Regulations).

You should have received an "original" Certificate and the contractor should have retained a duplicate. If you were the person ordering the work, but not the owner of the installation, you should pass this Certificate, or a full copy of it including the schedules, immediately to the owner.

The "original" Certificate should be retained in a safe place and be shown to any person inspecting or undertaking further work on the electrical installation in the future. If you later vacate the property, this Certificate will demonstrate to the new owner that the electrical installation complied with the requirements of British Standard 7671 at the

Appendix G

time the Certificate was issued. The Construction (Design and Management) Regulations require that, for a project covered by those Regulations, a copy of this Certificate, together with schedules, is included in the project health and safety documentation.

For safety reasons, the electrical installation will need to be inspected at appropriate intervals by a competent person. The maximum time interval recommended before the next inspection is stated on Page 1 under "NEXT INSPECTION".

This Certificate is intended to be issued only for a new electrical installation or for new work associated with an addition or alteration to an existing installation. It should not have been issued for the inspection of an existing electrical installation. An "Electrical Installation Condition Report" should be issued for such an inspection.

G4.4 Schedule of Test Results

Notes to the tests and observations required when completing the Schedule of Test Results:

- Measurement of Z_s at this distribution board to be recorded
- Measurement of I_{pf} at this distribution board to be recorded
- Confirm correct polarity of supply to this distribution board by the use of approved test instrument
- Confirmation of phase sequence for multi-phase installations
- Identify circuits with equipment which could be damaged if connected when tests are carried out, e.g. SELV transformers, dimming equipment.

The following tests, where relevant, must be carried out in the given sequence (see also 10.2):

A – Installation isolated from the supply

1   Continuity

Radial conductors

Continuity of protective conductors, including main and supplementary bonding

Every protective conductor, including main and supplementary bonding conductors, should be tested to verify that it is continuous and correctly connected.

Test method 1

Where test method 1 is used, enter the measured resistance of the line conductor plus the circuit protective conductor ($R_1 + R_2$). See 10.3.1. During the continuity testing (test method 1) the following polarity checks should be carried out:

1 overcurrent devices and single-pole controls are in the line conductor,
2 except for E14 and E27 lampholders to BS EN 60238, centre contact screw lampholders have the outer threaded contact connected to the neutral, and
3 socket-outlet polarities are correct.

Compliance for each circuit is indicated by a tick in polarity column 17.
($R_1 + R_2$) need not be recorded if R_2 is recorded in column 14.

G | Appendix

Test method 2
Where test method 2 is used, the maximum value of R_2 is recorded in column 14.

Ring final circuit continuity
Each conductor of the ring final circuit must be tested for continuity, including spurs. An exception is permitted where the cpc is formed by, e.g. metallic conduit or trunking and is not in the form of a ring. N/A can be recorded here but continuity of the cpc will be confirmed in either column 13 or 14.

2   Insulation resistance
All voltage sensitive devices to be disconnected or test between live conductors (line and neutral) connected together and earth.

The insulation resistance between live conductors (line-to-line and line-to-neutral for three-phase installations and line-to-neutral for single-phase installations) is inserted in column 15 and between live conductors and earth in column 16.

The minimum insulation resistance values are given in Table 10.3.3 of this Guide.

3   Polarity – by continuity method
A satisfactory polarity test may be indicated by a tick in column 17. Only in a Schedule of Test Results associated with an Electrical Installation Condition Report is it acceptable to record incorrect polarity.

B – Installation energised

4   Polarity of supply
The polarity of the supply at the distribution board should be confirmed and indicated by ticking the box on the Schedule of Test Results.

5   Earth fault loop impedance Zs
This may be determined either by direct measurement at the furthest point of a live circuit or by adding $(R_1 + R_2)$ of column 13 to Z_e. Z_e is determined by measurement at the origin of the installation.

$$Z_s = Z_e + (R_1 + R_2)$$

Z_s should not exceed the values given in Appendix B.

6   Functional testing
The operation of RCDs (including RCBOs) is tested by simulating a fault condition, independent of any test facility in the device; see Section 11.

When testing an RCD at IΔn, record the operating time in column 19.

Where RCDs rated at 30 mA or less are used to provide additional protection, the devices are also to be tested at 5IΔn and the operating time recorded in column 20.

Effectiveness of the test button must be confirmed and the result recorded in column 21.

Appendix G

7 Switchgear

All switchgear and controlgear assemblies, controls, etc. must be operated to ensure that they are properly mounted, adjusted and installed.

8 Earth electrode resistance

The resistance of earth electrodes must be measured. For reliability in service the resistance of any earth electrode should be below 200 Ω. Record the value on Form 1, 2 or 6, as appropriate.

G4.5 Minor Electrical Installation Works Certificate

Figure G4.5 shows an example of a completed Minor Electrical Installation Works Certificate and Table G4.8 gives some notes on how to complete it.

G | Appendix

▼ **Figure G4.5** Minor Electrical Installation Works Certificate – page 1 of 1

Form 5
Form No: .... SVT - /5

MINOR ELECTRICAL INSTALLATION WORKS CERTIFICATE
(REQUIREMENTS FOR ELECTRICAL INSTALLATIONS - BS 7671 [IET WIRING REGULATIONS])
To be used only for minor electrical work which does not include the provision of a new circuit

PART 1: Description of minor works

1. Description of the minor works **Addition of two socket-outlets, kitchen of dwelling.**
2. Location/Address **1 Latrigg Rise, Oldtown** Post Code **AC30 2BN**
3. Date minor works completed **4-Jan-2012**
4. Details of departures, if any, from BS 7671:2008, amended to ...**2011**... (date)
   None

PART 2: Installation details

1. System earthing arrangement   TN-C-S ☐   TN-S ☐   TT ☑
2. Method of fault protection **Automatic disconnection of supply (ADS)**
3. Protective device for the modified circuit   Type **BS EN 61009 Type C**   Rating ....**32**.. A

Comments on existing installation, including adequacy of earthing and bonding arrangements (see Regulation 132.16):
Earth leakage circuit-breaker (ELCB) used in utility area on older part of the installation; client advised that ELCB may not provide adequate protection and to have periodic inspection carried out on installation.
Earthing and bonding arrangements generally satisfactory.

PART 3: Essential Tests

Earth continuity satisfactory ☑

Insulation resistance:
    Line/neutral ....**+299**.. MΩ
    Line/earth ....**+299**.. MΩ
    Neutral/earth ....**+299**.. MΩ

Earth fault loop impedance ....**106**.. Ω

Polarity satisfactory ☑

RCD operation (if applicable). Rated residual operating current $I_{\Delta n}$ .....**30**...mA and operating time of ....**29**.ms (at $I_{\Delta n}$)

PART 4: Declaration

I/We CERTIFY that the said works do not impair the safety of the existing installation, that the said works have been designed, constructed, inspected and tested in accordance with BS 7671:2008 (IET Wiring Regulations), amended to **2011**........ (date) and that the said works, to the best of my/our knowledge and belief, at the time of my/our inspection, complied with BS 7671 except as detailed in Part 1 above.

Name: ....**Clive Jenkin**................. Signature: ....*Clive Jenkin*..........
For and on behalf of: **PCJ Electrical**............. Position: ....**Director**..........
Address: ....**22 Whinlatter Close**..........
                Oldtown Date: ....**4-Jan-2012**..........
                Post code.**AC30 8CD**

Page 1 of 1

Appendix | G

G4.6 Minor Electrical Installation Works Certificate – Scope of application

Notes:

1. The Minor Works Certificate is intended to be used for additions and alterations to an installation that do not extend to the provision of a new circuit. Examples include the addition of socket-outlets or lighting points to an existing circuit, the relocation of a light switch etc (see G2.2).
2. This Certificate may also be used for the replacement of equipment such as accessories or luminaires, but not for the replacement of distribution boards or similar items. Appropriate inspection and testing, however, should always be carried out irrespective of the extent of the work undertaken.

G4.7 Minor Electrical Installation Works Certificate – Guidance for recipients
(to be appended to the Certificate)

This Certificate has been issued to confirm that the electrical installation work to which it relates has been designed, constructed, inspected and tested in accordance with British Standard 7671 (the IET Wiring Regulations).

You should have received an "original" Certificate and the contractor should have retained a duplicate. If you were the person ordering the work, but not the owner of the installation, you should pass this Certificate, or a copy of it, to the owner. A separate Certificate should have been received for each existing circuit on which minor works have been carried out. This Certificate is not appropriate if you requested the contractor to undertake more extensive installation work, for which you should have received an Electrical Installation Certificate.

The Certificate should be retained in a safe place and be shown to any person inspecting or undertaking further work on the electrical installation in the future. If you later vacate the property, this Certificate will demonstrate to the new owner that the minor electrical installation work carried out complied with the requirements of British Standard 7671 at the time the Certificate was issued.

G | Appendix

G4.8 Notes on completion of the Minor Electrical Installation Works Certificate

▼ **Table G4.8** Description of the areas to be completed

Description of minor works	Information to record
1,2	The work to which the certificate applies must be so described that the work can be readily identified.
4	No departures are to be expected except in most unusual circumstances. See Regulations 120.3 and 133.5.
Part 2 – Installation details	
2	The method of fault protection must be clearly identified e.g. automatic disconnection of supply (ADS).
Comments on existing installation	The installer responsible for the new work should record on the Minor Electrical Installation Works Certificate any defects found, so far as is reasonably practicable, in the existing installation. The defects recorded should not affect the safety of the installation work to which the certificate applies.
Part 3 – Essential tests	The relevant provisions of Part 6 (Inspection and Testing) of BS 7671 must be applied in full to all minor works. For example, where a socket-outlet is added to an existing circuit it is necessary to: i) establish that the earth socket-tube of the socket-outlet is connected to the main earthing terminal ii) measure the insulation resistance of the circuit that has been added to and establish that it complies with Table 61 of BS 7671 iii) measure the earth fault loop impedance to establish that the maximum permitted disconnection time is not exceeded iv) check that the polarity of the socket-outlet is correct v) (if the work is protected by an RCD) verify the effectiveness of the RCD.
Part 4 – Declaration	
Regulation 631.4	The Certificate must be made out and signed by a competent person in respect of the design, construction, inspection and testing of the work.

Appendix G

▼ **Table G4.8** *continued*

Description of minor works	Information to record
	The competent person will have a sound knowledge and experience relevant to the nature of the work undertaken and to the technical standards set down in BS 7671, be fully versed in the inspection and testing procedures contained in the Regulations and employ adequate testing equipment.
	When making out and signing a form on behalf of a company or other business entity, individuals must state for whom they are acting.

G4.9 Electrical Installation Condition Report (EICR)

Installations may be divided into two types:

- ▶ Domestic and similar installations with up to 100 A single- or three-phase supply
- ▶ Installations with a supply greater than 100 A.

However, this Guide will only consider the Electrical Installation Condition Report for Domestic and similar installations with up to 100 A supply. For installations with a supply greater than 100 A, see IET Guidance Note 3.

For domestic and similar installations with up to 100 A supply, the inspector will be required to complete a minimum of five pages of information for an EICR.

An Electrical Installation Condition Report (Form 6) is to be issued for all inspected installations.

Figures G4.9(i)–(v) show a typical completed Electrical Installation Condition Report comprising Forms 6, 7 and 4. The installation is some 20 years old and has no RCD fitted.

G | Appendix

▼ **Figure G4.9(i)** Electrical Installation Condition Report – page 1

Form 6   Form No: S.V.T. -2 /6
ELECTRICAL INSTALLATION CONDITION REPORT

SECTION A. DETAILS OF THE CLIENT / PERSON ORDERING THE REPORT
Name ... J. Grassmoor
Address ... 43 Derwent Water
... Honister ... Post Code: AC10 3ER

SECTION B. REASON FOR PRODUCING THIS REPORT Client's request due to burning smell. Known rodent infestation, suspected cable damage in loft area.
Date(s) on which inspection and testing was carried out ...

SECTION C. DETAILS OF THE INSTALLATION WHICH IS THE SUBJECT OF THIS REPORT
Occupier ... J. Grassmoor
Address ... 43 Derwent Water
... Honister ... Post Code: AC10 3ER
Description of premises (tick as appropriate)
Domestic ☑ Commercial ☐ Industrial ☐ Other (include brief description) ☐
Estimated age of wiring system ... 20 ... years
Evidence of additions / alterations Yes ☑ No ☐ Not apparent ☐ If yes, estimate age ... 4 ... years
Installation records available? (Regulation 621.1) Yes ☐ No ☑ Date of last inspection ... Not known ... (date)

SECTION D. EXTENT AND LIMITATIONS OF INSPECTION AND TESTING
Extent of the electrical installation covered by this report
Visual inspection to distributor's equipment and electric meter, inspection and test of consumer unit and final circuits.
Agreed limitations including the reasons (see Regulation 634.2) No dismantling/removal of fitted kitchen units or appliances.
Agreed with: Client
Operational limitations including the reasons (see page no ....) None

The inspection and testing detailed in this report and accompanying schedules have been carried out in accordance with BS 7671: 2008 (IET Wiring Regulations) as amended to 2011
It should be noted that cables concealed within trunking and conduits, under floors, in roof spaces, and generally within the fabric of the building or underground, have **not** been inspected unless specifically agreed between the client and inspector prior to the inspection.

SECTION E. SUMMARY OF THE CONDITION OF THE INSTALLATION
General condition of the installation (in terms of electrical safety) Cable damage evident in loft, otherwise condition of the installation is generally good, although some signs of wear and tear.

Overall assessment of the installation in terms of its suitability for continued use
SATISFACTORY / UNSATISFACTORY* (Delete as appropriate)
*An unsatisfactory assessment indicates that dangerous (code C1) and/or potentially dangerous (code C2) conditions have been identified.

SECTION F. RECOMMENDATIONS
Where the overall assessment of the suitability of the installation for continued use above is stated as UNSATISFACTORY, I / we recommend that any observations classified as 'Danger present' (code C1) or 'Potentially dangerous' (code C2) are acted upon as a matter of urgency.
Investigation without delay is recommended for observations identified as 'further investigation required'.
Observations classified as 'Improvement recommended' (code C3) should be given due consideration.

Subject to the necessary remedial action being taken, I / we recommend that the installation is further inspected and tested by Jan-2015 (date)

SECTION G. DECLARATION
I/We, being the person(s) responsible for the inspection and testing of the electrical installation (as indicated by my/our signatures below), particulars of which are described above, having exercised reasonable skill and care when carrying out the inspection and testing, hereby declare that the information in this report, including the observations and the attached schedules, provides an accurate assessment of the condition of the electrical installation taking into account the stated extent and limitations in section D of this report.

Inspected and tested by:	Report authorised for issue by:
Name (Capitals) CLIVE JENKIN	Name (Capitals) CLIVE JENKIN
Signature	Signature
For/on behalf of PCJ Electrical	For/on behalf of PCJ Electrical
Position Director	Position Director
Address 22 Whinlatter Close, Oldtown	Address 22 Whinlatter Close, Oldtown
Post code AC30 8CD Date 4-Jan-2012	Post code AC30 8CD Date 4-Jan-2012

SECTION H. SCHEDULE(S)
... 1 ... schedule(s) of inspection and ... 1 ... schedule(s) of test results are attached.
The attached schedule(s) are part of this document and this report is valid only when they are attached to it.

Page 1 of 5.

Appendix G

▼ **Figure G4.9(ii)** Electrical Installation Condition Report – page 2

Form 6   Form No: SVT -2 ..../6

SECTION I. SUPPLY CHARACTERISTICS AND EARTHING ARRANGEMENTS

Earthing arrangements	Number and Type of Live Conductors	Nature of Supply Parameters	Supply Protective Device
TN-C ☐	a.c. ☑ d.c. ☐	Nominal voltage, U / $U_0^{(1)}$ ....230.. V	BS (EN) BS 1361
TN-S ☐	1-phase, 2-wire ☑ 2-wire ☐	Nominal frequency, $f^{(1)}$ ....50.. Hz	Type ...II...
TN-C-S ☑	1-phase, 3-wire ☐ 3-wire ☐	Prospective fault current, $I_{pf}^{(2)}$ ....1.4.. kA	Rated current ....80.... A
TT ☐	2 phase, 3-wire ☐ Other ☐	External loop impedance, $Z_e^{(2)}$ ....0.16.. Ω	
IT ☐	3 phase, 3-wire ☐	Note: (1) by enquiry	
	3 phase, 4-wire ☐	(2) by enquiry or by measurement	
	Confirmation of supply polarity ☑		

Other sources of supply (as detailed on attached schedule) ☐

SECTION J. PARTICULARS OF INSTALLATION REFERRED TO IN THE REPORT

Means of Earthing   **Details of Installation Earth Electrode** *(where applicable)*

Distributor's facility ☑   Type ......N/A..
Installation earth electrode ☐   Location ...N/A...
    Resistance to Earth ....N/A. Ω

Main Protective Conductors

Earthing conductor	Material Copper................	csa ....16......mm²	Connection / continuity verified ☑
Main protective bonding conductors	Material Copper................	csa ....10......mm²	Connection / continuity verified ☑
To incoming water service ☑	To incoming gas service ☐	To incoming oil service ☐	To structural steel ☐
To lightning protection ☐	To other incoming service(s) ☐ Specify N/A...		

Main Switch / Switch-Fuse / Circuit-Breaker / RCD

Location Garage........................	Current rating ....100. A	**If RCD main switch**	
....	Fuse / device rating or setting N/A A	Rated residual operating current ($I_{\Delta n}$) ....N/A..mA	
BS(EN) .BS 5486....................	Voltage rating ....250. V	Rated time delay ....N/A..ms	
No of poles 2.................................		Measured operating time (at $I_{\Delta n}$) ....N/A..ms	

SECTION K. OBSERVATIONS

Referring to the attached schedules of inspection and test results, and subject to the limitations specified at the *Extent and limitations of inspection and testing* section

No remedial action is required ☐   The following observations are made ☑ (see below):

OBSERVATION(S)	CLASSIFICATION CODE	FURTHER INVESTIGATION REQUIRED (YES / NO)
1. Damage to cable for shower circuit (Cir.No.7) in loft, conductors visible, arcing evident	C1	No
2. Damage to cable for lighting circuit (Cir.No.6) in loft, conductors visible	C1	No
3. No additional protection by RCD to socket-outlets used by ordinary persons	C3	No
4. No additional protection by RCD for equipment used outdoors	C3	No
5. No earthed, mechanical or additional protection for cables concealed in partition walls	C3	No
6. No additional protection for low voltage circuits in bathroom	C3	No

One of the following codes, as appropriate, has been allocated to each of the observations made above to indicate to the person(s) responsible for the installation the degree of urgency for remedial action.

C1 – Danger present. Risk of injury. Immediate remedial action required
C2 – Potentially dangerous - urgent remedial action required
C3 – Improvement recommended

Page 2 of .5.

G | Appendix

▼ **Figure G4.9(iii)** Electrical Installation Condition Report – page 3

Form 7   Form No: S.V.T.-2..../7

**CONDITION REPORT INSPECTION SCHEDULE FOR
DOMESTIC AND SIMILAR PREMISES WITH UP TO 100 A SUPPLY**
Note: This form is suitable for many types of smaller installation not exclusively domestic.

OUTCOMES	Acceptable condition	✓	Unacceptable condition	State C1 or C2	Improvement recommended	State C3	Not verified	N/V	Limitation	LIM	Not applicable	N/A

ITEM NO	DESCRIPTION	OUTCOME (Use codes above. Provide additional comment where appropriate. C1, C2 and C3 coded items to be recorded in Section K of the Condition Report)	Further investigation required? (Y or N)
1.0	**DISTRIBUTOR'S / SUPPLY INTAKE EQUIPMENT**		
1.1	Service cable condition	N/V	No
1.2	Condition of service head	✓	No
1.3	Condition of tails - Distributor	✓	No
1.4	Condition of tails - Consumer	✓	No
1.5	Condition of metering equipment	✓	No
1.6	Condition of isolator (where present)	N/A	No
2.0	**PRESENCE OF ADEQUATE ARRANGEMENTS FOR OTHER SOURCES SUCH AS MICROGENERATORS (551.6; 551.7)**	N/A	No
3.0	**EARTHING / BONDING ARRANGEMENTS (411.3; Chap 54)**		
3.1	Presence and condition of distributor's earthing arrangement (542.1.2.1; 542.1.2.2)	✓	No
3.2	Presence and condition of earth electrode connection where applicable (542.1.2.3)	N/A	No
3.3	Provision of earthing / bonding labels at all appropriate locations (514.13)	✓	No
3.4	Confirmation of earthing conductor size (542.3; 543.1.1)	✓	No
3.5	Accessibility and condition of earthing conductor at MET (543.3.2)	✓	No
3.6	Confirmation of main protective bonding conductor sizes (544.1)	✓	No
3.7	Condition and accessibility of main protective bonding conductor connections (543.3.2; 544.1.2)	✓	No
3.8	Accessibility and condition of all protective bonding connections (543.3.2)	✓	No
4.0	**CONSUMER UNIT(S) / DISTRIBUTION BOARD(S)**		
4.1	Adequacy of working space / accessibility to consumer unit / distribution board (132.12; 513.1)	✓	No
4.2	Security of fixing (134.1.1)	✓	No
4.3	Condition of enclosure(s) in terms of IP rating etc (416.2)	✓	No
4.4	Condition of enclosure(s) in terms of fire rating etc (526.5)	✓	No
4.5	Enclosure not damaged/deteriorated so as to impair safety (621.2(iii))	✓	No
4.6	Presence of main linked switch (as required by 537.1.4)	✓	No
4.7	Operation of main switch (functional check) (612.13.2)	✓	No
4.8	Manual operation of circuit-breakers and RCDs to prove disconnection (612.13.2)	✓	No
4.9	Correct identification of circuit details and protective devices (514.8.1; 514.9.1)	✓	No
4.10	Presence of RCD quarterly test notice at or near consumer unit / distribution board (514.12.2)	C3	No
4.11	Presence of non-standard (mixed) cable colour warning notice at or near consumer unit / distribution board (514.14)	✓	No
4.12	Presence of alternative supply warning notice at or near consumer unit / distribution board (514.15)	N/A	No
4.13	Presence of other required labelling (please specify) (Section 514)	N/A	No
4.14	Examination of protective device(s) and base(s); correct type and rating (no signs of unacceptable thermal damage, arcing or overheating) (421.1.3)	✓	No
4.15	Single-pole protective devices in line conductor only (132.14.1; 530.3.2)	✓	No
4.16	Protection against mechanical damage where cables enter consumer unit / distribution board (522.8.1; 522.8.11)	✓	No
4.17	Protection against electromagnetic effects where cables enter consumer unit / distribution board / enclosures (521.5.1)	✓	No
4.18	RCD(s) provided for fault protection – includes RCBOs (411.4.9; 411.5.2; 531.2)	C3	No
4.19	RCD(s) provided for additional protection - includes RCBOs (411.3.3; 415.1)	C3	No

Page .3. of .5.

Appendix G

▼ **Figure G4.9(iv)** Electrical Installation Condition Report – page 4

Form 7　　　　　　　　　　　　　　　　　　　　　Form No: SVT-2 ...../7

OUTCOMES	Acceptable condition	✓	Unacceptable condition	State C1 or C2	Improvement recommended	State C3	Not verified	N/V	Limitation	LIM	Not applicable	N/A

ITEM NO	DESCRIPTION	OUTCOME (Use codes above. Provide additional comment where appropriate. C1, C2 and C3 coded items to be recorded in Section K of the Condition Report)	Further investigation required? (Y or N)
5.0	**FINAL CIRCUITS**		
5.1	Identification of conductors (514.3.1)	✓	No
5.2	Cables correctly supported throughout their run (522.8.5)	✓	No
5.3	Condition of insulation of live parts (416.1)	C1	No
5.4	Non-sheathed cables protected by enclosure in conduit, ducting or trunking (521.10.1)	N/A	No
	• To include the integrity of conduit and trunking systems (metallic and plastic)	N/A	No
5.5	Adequacy of cables for current-carrying capacity with regard for the type and nature of installation (Section 523)	✓	No
5.6	Coordination between conductors and overload protective devices (433.1; 533.2.1)	✓	No
5.7	Adequacy of protective devices: type and rated current for fault protection (411.3)	✓	No
5.8	Presence and adequacy of circuit protective conductors (411.3.1.1; 543.1)	✓	No
5.9	Wiring system(s) appropriate for the type and nature of the installation and external influences (Section 522)	C1	No
5.10	Concealed cables installed in prescribed zones (see Section D. *Extent and limitations*) (522.6.101)	✓	No
5.11	Concealed cables incorporating earthed armour or sheath, or run within earthed wiring system, or otherwise protected against mechanical damage from nails, screws and the like (see Section D. *Extent and limitations*) (522.6.101; 522.6.103)	C3	No
5.12	Provision of additional protection by RCD not exceeding 30 mA:		
	• for all socket-outlets of rating 20 A or less provided for use by ordinary persons unless an exception is permitted (411.3.3)	C3	No
	• for supply to mobile equipment not exceeding 32 A rating for use outdoors (411.3.3)	C3	No
	• for cables concealed in walls or partitions (522.6.102; 522.6.103)	C3	No
5.13	Provision of fire barriers, sealing arrangements and protection against thermal effects (Section 527)	✓	No
5.14	Band II cables segregated / separated from Band I cables (528.1)	✓	No
5.15	Cables segregated / separated from communications cabling (528.2)	✓	No
5.16	Cables segregated / separated from non-electrical services (528.3)	✓	No
5.17	Termination of cables at enclosures – indicate extent of sampling in Section D of the report (Section 526)		
	• Connections soundly made and under no undue strain (526.6)	✓	No
	• No basic insulation of a conductor visible outside enclosure (526.8)	✓	No
	• Connections of live conductors adequately enclosed (526.5)	✓	No
	• Adequately connected at point of entry to enclosure (glands, bushes etc.) (522.8.5)	✓	No
5.18	Condition of accessories including socket-outlets, switches and joint boxes (621.2(iii))	✓	No
5.19	Suitability of accessories for external influences (512.2)	✓	No
6.0	**LOCATION(S) CONTAINING A BATH OR SHOWER**		
6.1	Additional protection for all low voltage (LV) circuits by RCD not exceeding 30 mA (701.411.3.3)	C3	No
6.2	Where used as a protective measure, requirements for SELV or PELV met (701.414.4.5)	N/A	No
6.3	Shaver sockets comply with BS EN 61558-2-5 formerly BS 3535 (701.512.3)	N/A	No
6.4	Presence of supplementary bonding conductors, unless not required by BS 7671:2008 (701.415.2)	✓	No
6.5	Low voltage (e.g. 230 volt) socket-outlets sited at least 3 m from zone 1 (701.512.3)	N/A	No
6.6	Suitability of equipment for external influences for installed location in terms of IP rating (701.512.2)	✓	No
6.7	Suitability of equipment for installation in a particular zone (701.512.3)	✓	No
6.8	Suitability of current-using equipment for particular position within the location (701.55)	✓	No
7.0	**OTHER PART 7 SPECIAL INSTALLATIONS OR LOCATIONS**		
7.1	List all other special installations or locations present, if any. (Record separately the results of particular inspections applied.)	N/A	No

Inspected by:
Name (Capitals) **CLIVE JENKIN** .... Signature *Clive Jenkin* .... Date 4-Jan-2012

Page .4. of .5.

G | Appendix

▶ **Figure G4.9(v)** Generic schedule of test results - Electrical Installation Condition Report – page 5

Form 4

Form No: SVT-2./4

GENERIC SCHEDULE OF TEST RESULTS

DB reference no	CU1
Location	Kitchen boiler cupboard
Zs at DB (Ω)	0.16
Ipf at DB (kA)	1.4
Correct supply polarity confirmed	☑
Phase sequence confirmed (where appropriate)	N/A

Details of circuits and/or installed equipment vulnerable to damage when testing: Dimming equipment in main lounge.

Details of test instruments used (state serial and/or asset numbers)	
Continuity	5463-7293
Insulation resistance	5463-7293
Earth fault loop impedance	5463-7293
RCD	5463-7293
Earth electrode resistance	N/A

Tested by:	
Name (Capitals)	CLIVE JENKIN
Signature	Clive Jenkin
Date	4-Jan-2012

Test results

Circuit number	Circuit Description	Circuit details — BS (EN)	Overcurrent device — type	Overcurrent device — rating (A)	Overcurrent device — breaking capacity (kA)	Conductor details — Reference Method	Conductor details — Live (mm²)	Conductor details — cpc (mm²)	Ring final circuit continuity (Ω) r_1 (line)	r_n (neutral)	r_2 (cpc)	Continuity (Ω) $(R_1 + R_2)$ *	R_2	Insulation Resistance (MΩ) Live-Live	Live-E	Polarity	Z_s (Ω)	RCD @ $I_{\Delta n}$ (ms)	@ 5$I_{\Delta n}$ (ms)	Test button / functionality	Remarks (continue on a separate sheet if necessary)
1	Ring - sockets downstairs	BS 3871	1	30	2	C	2.5	1.5	0.48	0.48	0.79	0.31	N/A	+299	+299	✓	0.48	N/A	N/A	N/A	
2	Ring - sockets upstairs	BS 3871	1	30	2	C	2.5	1.5	0.33	0.33	0.54	0.22	N/A	+299	+299	✓	0.38	N/A	N/A	N/A	
3	Ring - kitchen and utility	BS 3871	1	30	2	C	2.5	1.5	0.7	0.7	1.15	0.46	N/A	+299	+299	✓	0.62	N/A	N/A	N/A	
4	Lights - upstairs	BS 3871	1	5	2	C	1.0	1.0	N/A	N/A	N/A	3.07	N/A	+299	+299	✓	3.24	N/A	N/A	N/A	
5	Lights - downstairs and utility	BS 3871	1	5	2	C	1.0	1.0	N/A	N/A	N/A	3.8	N/A	+299	+299	✓	3.96	N/A	N/A	N/A	
6	Lights - garage	BS 3871	1	5	2	C	1.0	1.0	N/A	N/A	N/A	0.36	N/A	+299	+299	✓	0.52	N/A	N/A	N/A	
7	Shower (8 kW)	BS 3871	1	40	2	C	6.0	2.5	N/A	N/A	N/A	0.15	N/A	+299	+299	✓	0.31	N/A	N/A	N/A	
8	Spare																				

* Where there are no spurs connected to a ring final circuit this value is also the $(R_1 + R_2)$ of the circuit.

NOTE: One schedule of test results will be issued for every consumer unit or distribution board

Page 5 of 5

Appendix H
Standard circuit arrangements for household and similar installations

H1 Introduction

This appendix gives advice on standard circuit arrangements for household and similar premises. The circuits provide guidance on the requirements of Chapter 43 for overload protection and Section 537 of BS 7671 for isolation and switching. Reference must also be made to Section 7 and Table 7.1(i) for cable csa, length and installation reference method.

It is the responsibility of the designer and installer when adopting these circuit arrangements to take the appropriate measures to comply with the requirements of other chapters or sections which are relevant, such as Chapter 41 'Protection against electric shock', Chapter 54 'Earthing arrangements and protective conductors' and Chapter 52 'Selection and erection of wiring systems'.

Circuit arrangements other than those detailed in this appendix are not precluded when specified by a competent person, in accordance with the general requirements of Regulation 314.3.

H2 Final circuits using socket-outlets complying with BS 1363-2 and fused connection units complying with BS 1363-4

H2.1 General

In this arrangement, a ring or radial circuit, with spurs if any, feeds permanently connected equipment and a number of socket-outlets and fused connection units.

The floor area served by the circuit is determined by the known or estimated load and should not exceed the value given in Table H2.1.

H | Appendix

433.1.103 A single 30 A or 32 A ring circuit may serve a floor area of up to 100 m². Socket-outlets for washing machines, tumble dryers and dishwashers should be located so as to provide reasonable sharing of the load in each leg of the ring, or consideration should be given to separate circuits.

553.1.7 The number of socket-outlets provided should be such that all equipment can be supplied from an adjacent accessible socket-outlet, taking account of the length of flex normally fitted to portable appliances and luminaires.

Diversity between socket-outlets and permanently connected equipment has already been taken into account in Table H2.1 and no further diversity should be applied, see Appendix A of this Guide.

▼ **Table H2.1** Final circuits using BS 1363 socket-outlets and connection units

			Minimum live conductor cross-sectional area* (mm²)		
Type of Circuit		Overcurrent protective device rating (A)	Copper conductor thermoplastic or thermosetting insulated cables	Copper conductor mineral insulated cables	Maximum floor area served (m²)
1	2	3	4	5	6
A1	Ring	30 or 32	2.5	1.5	100
A2	Radial	30 or 32	4	2.5	75
A3	Radial	20	2.5	1.5	50

* See Section 7 and Table 7.1(i) for the minimum csa for particular installation reference methods. It is permitted to reduce the values of conductor cross-sectional area for fused spurs.

Where two or more ring final circuits are installed, the socket-outlets and permanently connected equipment to be served should be reasonably distributed among the circuits.

H2.2 Circuit protection

Table H2.1 is applicable for circuits protected by:

- ▶ fuses to BS 3036, BS 1361 and BS 88, and
- ▶ circuit-breakers:
  - Types B and C to BS EN 60898 or BS EN 61009-1
  - BS EN 60947-2
  - Types 1, 2 and 3 to BS 3871.

Appendix **H**

H2.3 Conductor size

The minimum size of conductor cross-sectional area in the circuit and in non-fused spurs is given in Table H2.1, however, the actual size of cable is determined by the current-carrying capacity for the particular method of installation, after applying appropriate rating factors from Appendix F, see Table 7.1(i). The as-installed current-carrying capacity (I_z) so calculated must be not less than:

- 20 A for ring circuit A1
- 30 A or 32 A for radial circuit A2 (i.e. the rating of the overcurrent protective device)
- 20 A for radial circuit A3 (i.e. the rating of the overcurrent protective device).

The conductor size for a fused spur is determined from the total current demand served by that spur, which is limited to a maximum of 13 A.

Where a fused spur serves socket-outlets the minimum conductor size is:

- 1.5 mm² for cables with thermosetting or thermoplastic (PVC) insulated cables, copper conductors
- 1 mm² for mineral insulated cables, copper conductors.

The conductor size for circuits protected by BS 3036 fuses is determined by applying the 0.725 factor of Regulation 433.1.101, that is the current-carrying capacity must be at least 27 A for circuits A1 and A3, 41 A for circuit A2.

H2.4 Spurs

The total number of fused spurs is unlimited but the number of non-fused spurs should not exceed the total number of socket-outlets and items of stationary equipment connected directly in the circuit.

In an A1 ring final circuit and an A2 radial circuit of Table H2.1 a non-fused spur should feed only one single or one twin or multiple socket-outlet or one item of permanently connected equipment. Such a spur should be connected to the circuit at the terminals of a socket-outlet or junction box, or at the origin of the circuit in the distribution board.

A fused spur should be connected to the circuit through a fused connection unit, the rating of the fuse in the unit not exceeding that of the cable forming the spur and, in any event, not exceeding 13 A.

H2.5 Permanently connected equipment

Permanently connected equipment should be locally protected by a fuse complying with BS 1362 of rating not exceeding 13 A or by a circuit-breaker of rating not exceeding 16 A and should be controlled by a switch, where needed (see Appendix J). A separate switch is not required if the circuit-breaker is to be used as a switch.

H3 Radial final circuits using 16 A socket-outlets complying with BS EN 60309-2 (BS 4343)

H3.1 General
Where a radial circuit feeds equipment the maximum demand of which, having allowed for diversity, is known or estimated not to exceed the rating of the overcurrent protective device and in any event does not exceed 20 A, the number of socket-outlets is unlimited.

H3.2 Circuit protection
The overcurrent protective device should have a rating not exceeding 20 A.

H3.3 Conductor size
The minimum size of conductor in the circuit is given in Tables H2.1 and 7.1(i). Where cables are grouped together the limitations of 7.2.1 and Appendix F apply.

H3.4 Types of socket-outlet
Socket-outlets should have a rated current of 16 A and be of the type appropriate to the number of phases, circuit voltage and earthing arrangements. Socket-outlets incorporating pilot contacts are not included.

H4 Cooker circuits in household and similar premises

The circuit supplies a control switch or a cooker unit complying with BS 4177, which may incorporate a socket-outlet.

The rating of the circuit is determined by the assessment of the current demand of the cooking appliance(s), and cooker control unit socket-outlet if any, in accordance with Table A1 of Appendix A. A 30 or 32 A circuit is usually appropriate for household or similar cookers of rating up to 15 kW.

A circuit of rating exceeding 15 A but not exceeding 50 A may supply two or more cooking appliances where these are installed in one room. The control switch or cooker control unit should be placed within 2 m of the appliance, but not directly above it. Where two stationary cooking appliances are installed in one room, one switch may be used to control both appliances provided that neither appliance is more than 2 m from the switch. Attention is drawn to the need to provide selective (discriminative) operation of protective devices as stated in Regulation 536.2.

Appendix H

H5 Water and space heating

Water heaters fitted to storage vessels in excess of 15 litres capacity, or permanently connected heating appliances forming part of a comprehensive space heating installation, should be supplied by their own separate circuit.

Immersion heaters should be supplied through a switched cord-outlet connection unit complying with BS 1363-4.

H6 Height of switches, socket-outlets and controls

553.1.6

The Building Regulations of England and Wales and of Scotland require switches and socket-outlets in new dwellings to be installed so that all persons including those whose reach is limited can easily use them. A way of satisfying the requirement is to install switches, socket-outlets and controls throughout the dwelling in accessible positions at a height of between 450 mm and 1200 mm from the finished floor level – see Figure H6. Because of the sensitivity of circuit-breakers, RCCBs and RCBOs fitted in consumer units, consumer units should be readily accessible.

(In areas subject to flooding, meters, cut-outs and consumer units should preferably be fixed above flood water level.)

▼ **Figure H6** Height of switches, socket-outlets, etc.

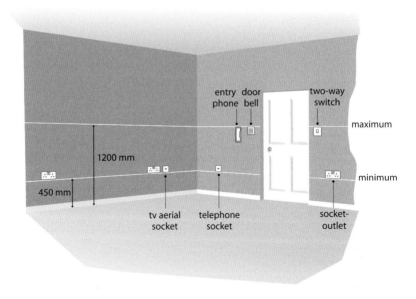

H | Appendix

H7 Number of socket-outlets

553.1.7 Sufficient socket-outlets are required to be installed so that all equipment likely to be used can be supplied from a reasonably accessible socket-outlet, taking account of the length of flexible cable normally fitted to portable appliances and luminaires. Table H7 provides guidance on the number of socket-outlets that are likely to meet this requirement.

In Scotland, mandatory standard 4.6 requires that every building must be designed and constructed in such a way that electric lighting points and socket-outlets are provided to ensure the health, safety and convenience of occupants and visitors. The Building Standards Division of the Scottish Government make recommendations for the number of socket-outlets that should be installed in a domestic premises in section 4.6.4 of the domestic technical handbook as follows:

- kitchen – 6 (at least 3 above worktop height)
- other habitable rooms – 4
- plus at least 4 more throughout the property including at least one per circulation area per storey.

The socket-outlets may be either single or double.

▼ **Table H7** Minimum number of twin socket-outlets to be provided in homes

Room type	Smaller rooms (up to 12 m²)	Medium rooms (12–25 m²)	Larger rooms (more than 25 m²)
Main living room (note 4)	4	6	8
Dining room	3	4	5
Single bedroom (note 3)	2	3	4
Double bedroom (note 3)	3	4	5
Bedsitting room (note 6)	4	5	6
Study	4	5	6
Utility room	3	4	5
Kitchen (note 1)	6	8	10
Garage (note 2)	2	3	4
Conservatory	3	4	5
Hallway	1	2	3
Loft	1	2	3
Location containing a bath or shower		note 5	

Appendix H

Notes to Table H7:

1. KITCHEN – If a socket-outlet is provided in the cooker control unit, this should not be included in the 6 recommended in the table above.

   Appliances built into kitchen furniture (integrated appliances) should be connected to a socket-outlet or switch fused connection unit that is accessible when the appliance is in place and in normal use. Alternatively, where an appliance is supplied from a socket-outlet or a connection unit, this should be controlled by an accessible double-pole switch or switched fused connection unit.

   It is recommended that wall mounted socket-outlets above a work surface are spaced at not more than 1 m intervals along the surface.

2. GARAGE – The number of socket-outlets specified allows for the use of a battery charger, tools, portable light and garden appliances.

3. BEDROOM – It is envisaged that this room will be used in different ways in different households. It may be used simply as a child's bedroom requiring socket-outlets for table lamps, an electric blanket and an electric heater only; or it may serve as a teenager's bedroom and living room combined, where friends are entertained. In this case, socket-outlets may be needed for computers (printers, scanners, speakers, etc.), games consoles, MP3/4 players, mobile phone chargers, DVD players, digital receivers, home entertainment systems (amplifier, CD player), hairdryer, television and radio, in addition to lamps, an electric blanket and electric heater.

4. HOME ENTERTAINMENT – In addition to the number of socket-outlets shown in the table it is recommended that at least two further double socket-outlets are installed in home entertainment areas.

5. LOCATIONS CONTAINING A BATH OR SHOWER – Except for SELV socket-outlets complying with Section 414 and shaver supply units complying with BS EN 61558-2-5, socket-outlets are prohibited within a distance of 3 m horizontally from the boundary of zone 1.

6. BEDSITTING ROOM – Rooms specifically designed or envisaged to be used as student bedsitting rooms should be provided with additional socket-outlets which may be needed since persons using these rooms will often introduce other portable appliances in addition to items already mentioned in Note 3. In such situations a lack of sufficient socket-outlets is an additional danger and therefore the minimum number of twin outlets should be increased to four.

Appendix I
Resistance of copper and aluminium conductors

434.5.2
543.1.3

To check compliance with Regulation 434.5.2 and/or Regulation 543.1.3, i.e. to evaluate the equation $S^2 = I^2 \cdot t/k^2$, it is necessary to establish the impedances of the circuit conductors to determine the fault current I and hence the protective device disconnection time t.

Fault current $I = U_0/Z_s$

where:

U_0 is the nominal voltage to earth
Z_s is the earth fault loop impedance

and

$Z_s = Z_e + (R_1 + R_2)$

where:

Z_e is that part of the earth fault loop impedance external to the circuit concerned
R_1 is the resistance of the line conductor from the origin of the circuit to the point of utilization
R_2 is the resistance of the protective conductor from the origin of the circuit to the point of utilization.

Similarly, in order to design circuits for compliance with BS 7671 limiting values of earth fault loop impedance given in Tables 41.2 to 41.4, it is necessary to establish the relevant impedances of the circuit conductors concerned at their operating temperature.

Table I1 gives values of $(R_1 + R_2)$ per metre for various combinations of conductors up to and including 35 mm² cross-sectional area. It also gives values of resistance (milliohms) per metre for each size of conductor. These values are at 20 °C.

Appendix

▼ **Table I1** Values of resistance/metre or $(R_1 + R_2)$/metre for copper and aluminium conductors at 20 °C

Cross-sectional area (mm²)		Resistance/metre or $(R_1 + R_2)$/metre (mΩ/m)	
Line conductor	Protective conductor	Copper	Aluminium
1	–	18.10	
1	1	36.20	
1.5	–	12.10	
1.5	1	30.20	
1.5	1.5	24.20	
2.5	–	7.41	
2.5	1	25.51	
2.5	1.5	19.51	
2.5	2.5	14.82	
4	–	4.61	
4	1.5	16.71	
4	2.5	12.02	
4	4	9.22	
6	–	3.08	
6	2.5	10.49	
6	4	7.69	
6	6	6.16	
10	–	1.83	
10	4	6.44	
10	6	4.91	
10	10	3.66	
16	–	1.15	1.91
16	6	4.23	–
16	10	2.98	–
16	16	2.30	3.82
25	–	0.727	1.20
25	10	2.557	–
25	16	1.877	–
25	25	1.454	2.40
35	–	0.524	0.87
35	16	1.674	2.78
35	25	1.251	2.07
35	35	1.048	1.74
50	–	0.387	0.64
50	25	1.114	1.84
50	35	0.911	1.51
50	50	0.774	1.28

Appendix I

▼ **Table I2** Ambient temperature multipliers to Table I1

Expected ambient temperature (°C)	Correction factor*
5	0.94
10	0.96
15	0.98
20	1.00
25	1.02

* The correction factor is given by {1 + 0.004(ambient temp − 20 °C)} where 0.004 is the simplified resistance coefficient per °C at 20 °C given by BS EN 60228 for copper and aluminium conductors.

Verification

For verification purposes the designer will need to give the values of the line and circuit protective conductor resistances at the ambient temperature expected during the tests. This may be different from the reference temperature of 20 °C used for Table I1. The rating factors in Table I2 may be applied to the values to take account of the ambient temperature (for test purposes only).

Multipliers for conductor operating temperature

Table I3 gives the multipliers to be applied to the values given in Table I1 for the purpose of calculating the resistance at maximum operating temperature of the line conductors and/or circuit protective conductors in order to determine compliance with, as applicable, the earth fault loop impedance of Table 41.2, 41.3 or 41.4 of BS 7671.

Table 41.2
Table 41.3
Table 41.4

Where it is known that the actual operating temperature under normal load is less than the maximum permissible value for the type of cable insulation concerned (as given in the tables of current-carrying capacity) the multipliers given in Table I3 may be reduced accordingly.

Appendix

▼ **Table I3** Multipliers to be applied to Table I1 to calculate conductor resistance at maximum operating temperature (note 3) for standard devices (note 4)

Conductor installation	Conductor insulation		
	70 °C Thermoplastic (PVC)	90 °C Thermoplastic (PVC)	90 °C Thermosetting
Not incorporated in a cable and not bunched (note 1)	1.04	1.04	1.04
Incorporated in a cable or bunched (note 2)	1.20	1.28	1.28

Notes:

Table 54.2 **1** See Table 54.2 of BS 7671, which applies where the protective conductor is not incorporated or bunched with cables, or for bare protective conductors in contact with cable covering.

Table 54.3 **2** See Table 54.3 of BS 7671, which applies where the protective conductor is a core in a cable or is bunched with cables.

3 The multipliers given in Table I3 for both copper and aluminium conductors are based on a simplification of the formula given in BS EN 60228, namely that the resistance–temperature coefficient is 0.004 per °C at 20 °C.

4 Standard devices are those described in Appendix 3 of BS 7671 (fuses to BS 1361, BS 88, BS 3036, circuit-breakers to BS EN 60898 types B, C, and D) and BS 3871-1.

Appendix J

Selection of devices for isolation and switching

Table 53.4

▼ **Table J1** Summary of the functions provided by devices for isolation and switching

Device	Standard	Isolation[5]	Emergency switching[2,5]	Functional switching[5]
Switching device	BS EN 50428	No	No	Yes
	BS EN 60669-1	No	Yes	Yes
	BS EN 60669-2-1	No	No	Yes
	BS EN 60669-2-2	No	Yes	Yes
	BS EN 60669-2-3	No	Yes	Yes
	BS EN 60669-2-4	Yes	Yes	Yes
	BS EN 60947-3	Yes[1]	Yes	Yes
	BS EN 60947-5-1	No	Yes	Yes
Contactor	BS EN 60947-4-1	Yes[1]	Yes	Yes
	BS EN 61095	No	No	Yes
Circuit-breaker	BS EN 60898	Yes	Yes	Yes
	BS EN 60947-2	Yes[1]	Yes	Yes
	BS EN 61009-1	Yes	Yes	Yes
RCD	BS EN 60947-2	Yes[1]	Yes	Yes
	BS EN 61008-1	Yes	Yes	Yes
	BS EN 61009-1	Yes	Yes	Yes
Isolating switch	BS EN 60669-2-4	Yes	Yes	Yes
	BS EN 60947-3	Yes	Yes	Yes
Plug and socket-outlet ($\leq$ 32 A)	BS EN 60309	Yes	No	Yes
	IEC 60884	Yes	No	Yes
	IEC 60906	Yes	No	Yes
Plug and socket-outlet (> 32 A)	BS EN 60309	Yes	No	No

J | Appendix

▼ **Table J1** *continued*

Device	Standard	Isolation[5]	Emergency switching[2,5]	Functional switching[5]
Device for the connection of luminaire	BS EN 61995-1	Yes[3]	No	No
Control and protective switching device for equipment (CPS)	BS EN 60947-6-1 BS EN 60947-6-2	Yes[1] Yes[1]	Yes Yes	Yes Yes
Fuse	BS 88	Yes	No	No
Device with semiconductors	BS EN 60669-2-1	No	No	Yes
Luminaire-supporting coupler	BS 6972	Yes[3]	No	No
Plug and unswitched socket-outlet	BS 1363-1 BS 1363-2	Yes[3] Yes[3]	No No	Yes Yes
Plug and switched socket-outlet	BS 1363-1 BS 1363-2	Yes[3] Yes[3]	No No	Yes Yes
Plug and socket-outlet	BS 5733	Yes[3]	No	Yes
Switched fused connection unit	BS 1363-4	Yes[3]	Yes	Yes
Unswitched fused connection unit	BS 1363-4	Yes[3] (removal of fuse link)	No	No
Fuse	BS 1362	Yes	No	No
Cooker control unit switch	BS 4177	Yes	Yes	Yes

Notes:
1. Function provided if the device is suitable and marked with the symbol for isolation (see BS EN 60617 identity number S00288). ─/─
2. The means of operation shall be readily accessible at places where a danger might occur and, where appropriate, at any additional remote position from which that danger can be removed. *(537.4.2.5)*
3. Device is suitable for on-load isolation, i.e. disconnection whilst carrying load current.
4. In an installation forming part of a TT or IT system, isolation requires disconnection of all the live conductors. *(537.2.2.1)*
5. 'Yes' indicates function provided; 'No' indicates function not provided.

Appendix K
Identification of conductors

K1 Introduction

The requirements of BS 7671 were harmonized with the technical intent of CENELEC Standard HD 384.5.514: *Identification*, including 514.3: *Identification of conductors* (now withdrawn).

Amendment No 2:2004 (AMD 14905) to BS 7671 implemented the harmonized cable core colours and the alphanumeric marking of the following standards:

- ▶ HD 308 S2:2001 *Identification of cores in cables and flexible cords*
- ▶ BS EN 60445:2000 *Basic and safety principles for man–machine interface, marking and identification of equipment and terminals and of terminations*
- ▶ BS EN 60446:2000 *Basic and safety principles for man–machine interface, marking and identification of equipment by colours or numerals*

This appendix provides guidance on marking at the interface between old and harmonized colours, and general guidance on the colours to be used for conductors.

British Standards for fixed and flexible cables have been harmonized (see Table K1). BS 7671 has been modified to align with these cables but also allows other suitable methods of marking connections by colours, e.g. tapes, sleeves or discs, or by alphanumerics, i.e. letters and/or numbers. Methods may be mixed within an installation.

Appendix

▼ **Table K1** Identification of conductors (Harmonized)

Function	Alphanumeric	Colour
Protective conductors		Green-and-Yellow
Functional earthing conductor		Cream
a.c. power circuit[1]		
Line of single-phase circuit	L	Brown
Neutral of single- or three-phase circuit	N	Blue
Line 1 of three-phase a.c. circuit	L1	Brown
Line 2 of three-phase a.c. circuit	L2	Black
Line 3 of three-phase a.c. circuit	L3	Grey
Two-wire unearthed d.c. power circuit		
Positive of two-wire circuit	L+	Brown
Negative of two-wire circuit	L-	Grey
Two-wire earthed d.c. power circuit		
Positive (of negative earthed) circuit	L+	Brown
Negative (of negative earthed) circuit[2]	M	Blue
Positive (of positive earthed) circuit[2]	M	Blue
Negative (of positive earthed) circuit	L-	Grey
Three-wire d.c. power circuit		
Outer positive of two-wire circuit derived from three-wire system	L+	Brown
Outer negative of two-wire circuit derived from three-wire system	L-	Grey
Positive of three-wire circuit	L+	Brown
Mid-wire of three-wire circuit[2,3]	M	Blue
Negative of three-wire circuit	L-	Grey
Control circuits, ELV and other applications		
Line conductor	L	Brown, Black, Red, Orange, Yellow, Violet, Grey, White, Pink or Turquoise
Neutral or mid-wire[4]	N or M	Blue

Notes:
1. Power circuits include lighting circuits.
2. M identifies either the mid-wire of a three-wire d.c. circuit, or the earthed conductor of a two-wire earthed d.c. circuit.
3. Only the middle wire of three-wire circuits may be earthed.
4. An earthed PELV conductor is blue.

Appendix **K**

K2 Addition or alteration to an existing installation

K2.1 Single-phase

An addition or alteration made to a single-phase installation need not be marked at the interface provided that:

    i the old cables are correctly identified by the colours red for line and black for neutral, and

    ii the new cables are correctly identified by the colours brown for line and blue for neutral.

K2.2 Two- or three-phase installation

Where an addition or alteration is made to a two- or a three-phase installation wired in the old core colours with cable to the new core colours, unambiguous identification is required at the interface. Cores shall be marked as follows:

    Neutral conductors
    Old and new conductors: N

    Line conductors
    Old and new conductors: L1, L2, L3

Table 7A ▼ **Table K2** Example of conductor marking at the interface for additions and alterations to an a.c. installation identified with the old cable colours

Function	Old conductor		New conductor	
	Colour	Marking	Marking	Colour
Line 1 of a.c.	Red	L1	L1	Brown*
Line 2 of a.c.	Yellow	L2	L2	Black*
Line 3 of a.c.	Blue	L3	L3	Grey*
Neutral of a.c.	Black	N	N	Blue
Protective conductor	Green-and-Yellow			Green-and-Yellow

* Three single-core cables with insulation of the same colour may be used if identified at the terminations.

K3 Switch wires in a new installation or an addition or alteration to an existing installation

Where a two-core cable with cores coloured brown and blue is used as a switch wire, both conductors being line conductors, the blue conductor should be marked brown or L at its terminations.

K Appendix

K4 Intermediate and two-way switch wires in a new installation or an addition or alteration to an existing installation

Where a three-core cable with cores coloured brown, black and grey is used as a switch wire, all three conductors being line conductors, the black and grey conductors should be marked brown or L at their terminations.

K5 Line conductors in a new installation or an addition or alteration to an existing installation

Power circuit line conductors should be coloured as in Table K1. Other line conductors may be brown, black, red, orange, yellow, violet, grey, white, pink or turquoise.

In a two- or three-phase power circuit, the line conductors may all be of one of the permitted colours, either identified L1, L2, L3 or marked brown, black, grey at their terminations.

K6 Changes to cable core colour identification

Table 7B ▼ **Table K6(i)** Cable to BS 6004 (flat cable with bare cpc)

Cable type	Old core colours	New core colours
Single-core + bare cpc	Red or Black	Brown or Blue
Two-core + bare cpc	Red, Black	Brown, Blue
Alt. two-core + bare cpc	Red, Red	Brown, Brown
Three-core + bare cpc	Red, Yellow, Blue	Brown, Black, Grey

Table 7C ▼ **Table K6(ii)** Standard 600/1000 V armoured cable BS 6346, BS 5467 or BS 6724

Cable type	Old core colours	New core colours
Single-core	Red or Black	Brown or Blue
Two-core	Red, Black	Brown, Blue
Three-core	Red, Yellow, Blue	Brown, Black, Grey
Four-core	Red, Yellow, Blue, Black	Brown, Black, Grey, Blue
Five-core	Red, Yellow, Blue, Black, Green-and-Yellow	Brown, Black, Grey, Blue, Green-and-Yellow

Appendix K

Table 7D ▼ **Table K6(iii)** Flexible cable to BS 6500

Cable type	Old core colours	New core colours
Two-core	Brown, Blue	No change
Three-core	Brown, Blue, Green-and-Yellow	No change
Four-core	Black, Blue, Brown, Green-and-Yellow	Brown, Black, Grey, Green-and-Yellow
Five-core	Black, Blue, Brown, Black, Green-and-Yellow	Brown, Black, Grey, Blue, Green-and-Yellow

K7 Addition or alteration to a d.c. installation

Where an addition or alteration is made to a d.c. installation wired in the old core colours with cable to the new core colours, unambiguous identification is required at the interface. Cores should be marked as follows:

Neutral and midpoint conductors
Old and new conductors: M

Line conductors
Old and new conductors: Brown or Grey, or L+ or L−

Table 7E ▼ **Table K7** Example of conductor marking at the interface for additions and alterations to a d.c. installation identified with the old cable colours

Function	Old conductor Colour	Old conductor Marking	New conductor Marking	New conductor Colour
Two-wire unearthed d.c. power circuit				
Positive of two-wire circuit	Red	L+	L+	Brown
Negative of two-wire circuit	Black	L−	L−	Grey
Two-wire earthed d.c. power circuit				
Positive (of negative earthed) circuit	Red	L+	L+	Brown
Negative (of negative earthed) circuit	Black	M	M	Blue
Positive (of positive earthed) circuit	Black	M	M	Blue
Negative (of positive earthed) circuit	Blue	L−	L−	Grey
Three-wire d.c. power circuit				
Outer positive of two-wire circuit derived from three-wire system	Red	L+	L+	Brown
Outer negative of two-wire circuit derived from three-wire system	Red	L−	L−	Grey
Positive of three-wire circuit	Red	L+	L+	Brown
Mid-wire of three-wire circuit	Black	M	M	Blue
Negative of three-wire circuit	Blue	L−	L−	Grey

Index

A

Additional protection
    inspection   9.2.2viiid
    labelling   6.1
    provision by RCD   3.4.1.1
    supplementary bonding   4.7
    testing   11.5
Alphanumeric identification of
    conductors   Table K1
Alternative supplies, warning notice   6.14
Automatic disconnection (ADS)   3.4.1; 3.5; 9.2.2c

B

Bands I and II, segregation   7.4.1
Basic protection   3.4.1.1
Bath/shower   8
    cubicle not in bathroom   8.2
    general   3.6.1iii; 3.6.3; 7.2.5iii; 8; Table 3.4.3
    summary of requirements   8.1; Table 8.1
    underfloor heating   8.3.1
    zone diagrams   Figs 8.1(i)–(iii)
Bending radii of cables   Table D5
Bonding   4
BS1363 socket-outlets   Appx H
Building logbook   Foreword
Building Regulations   1.2

C

Cable
    communications   7.4.2
    floors and ceilings   7.3.1
    grouping   7.2.1
    in thermal insulation   Table 7.1(iii)
    lengths, maximum   Table 7.1(i)
    ratings   Table 7.1(ii); Appx F
    selection   Appx C
    separation distances   Tables 7.4.2(i), (ii)
    supports/bends   Appx D
    walls and partitions   7.3.2
Capacities
    conduit   Appx E
    trunking   Appx E
Ceilings, cables above   7.3.1; Tables 7.1(ii), (iii)
Certificates   9.1; Appx G
Charge retention, warning label   6.2
Circuit protective conductors   10.3.1; Appx B
    continuity test   10.3.1i
Circuit-breakers
    application   Table 7.2.7(ii)
    short-circuit capacity   Table 7.2.7(i)
Class I and Class II equipment   2.4.1
Colours, cable core   Appx K
Communications cables   7.4.2
Competent persons   Preface; Foreword
Conductor cross-sectional area   Table 7.1(ii)
Conduit
    capacities   Appx E
    supports   Table D3
Consumer unit   2.2.5; 3.3
    split   Figs 3.6.3(i)/(ii)/(iv)/(v)
    with RCBOs   Fig 3.6.3(iii)
Controlgear   2.2.5; 6.3
Cooker circuit   Appx H
Corrosion of cable   Appx C

Index

Current-carrying capacity	Appx F
Cut-out,	
distributor's	1.1iii; 1.3v; 2.2.1

D

Devices, selection of	
isolation and switching	Appx J
protective	7.2.7
Diagrams	6.11
Disconnection times	3.5; 7.1viii; 7.2.7iv; Appx B
Distribution board	3.1; 6.15
Distributor (definition)	1.1
Diversity	Appx A

E

Earth electrode	4.9; 4.10
testing	10.3.5; Fig 10.3.5.2
Earth fault loop impedance	1.1iv; 1.3iv; 3.6.1i; 7.2.5; 7.2.6; Table 7.1(i) note 1; Appx B
testing	9.3.1; 10.3.6
Earthing and bonding	4
conductor size	4.4
equipotential bonding, supplementary	4.6–4.8; Table 4.6
gas service pipe	4.4
generator reference	2.4.3
high protective conductor current	7.5
label	Fig 6.5
oil service pipe	4.4
TN-C-S	Fig 2.1(i); Table 4.4(i)
TN-S	Fig 2.1(ii); Table 4.4(i)
TT	Fig 2.1(iii); Table 4.4(ii)
typical arrangements	4.11
water service pipe	4.4
Electric shock, protection against	3.4; 8.1
Electrical Installation Certificates	9.1; Appx G
Electricity at Work Regulations	10.1
Emergency switching	5.3; Table J1
Energy-efficient lighting	1.2.1
External cables and telecommunications	Table 7.4.2(i)

F

Fault current	
prospective	1.3iii; 7.2.7
protection	3.3
Fault protection	3.4.1.2
FELV	10.3.3vi; Table 10.3.3
Final circuits	7
standard	7.2; Appx H
Fire safety requirements	1.2.1
Firefighter's switch	5.5
Flashover	3.7.1
Floating earth (portable generator)	2.4.1; Fig 2.4.1
Floors	7.3.1
Functional switching	5.4; Table J1
Functional testing	10.3.9
Furniture with electrical supply	7.6
Fuses	2.2.5; Table 7.2.7(i)
distributor's	1.1iii; 2.2.1

G

Garages	1.1a
Gas installations	2.3; 4.3; 4.4; 7.4.3
Generators, portable	2.4; Figs 2.4.1/ 2.4.2/2.4.3(i), (ii)

H

Height of overhead wiring	Table D2
Height of switches, socket-outlets	Appx H
High protective conductor current	
earthing	7.5
labelling at DB	Fig 6.15
HSE Guidance Note GS 38	10.1

I

Identification of conductors	Appx K
Immersed equipment	Table 3.4.3
Immersion heaters	Appx H
Induction loops (hearing)	7.4.4
Information	1.3
Initial testing	10
Inspection and testing	9
checklists	9.2.2; 9.3.1
label for periodic	Fig 6.10
report	Appx G
schedules	Appx G
Installation	
considerations	7.3
diagram	6.11
method	7.1
Insulation resistance	9.3.1; 10.3.3
minimum values	Table 10.3.3

Index

Internal cables and
  telecommunications Table 7.4.2(ii)
Isolation 5
  identification 6.8
  multiple devices 6.9
  requirements 5.1.1
  switch 2.2.3; 2.2.4; 2.2.5
  switchgear 5.1.2; Table J1

J
Joists/ceilings 7.3.1; Fig 7.3.1; Tables 7.1(ii), (iii)

L
Labelling 6
Lighting circuits Table 7.1(i); 7.2.3; 7.4.4
Load estimation Appx A
Logbook, building Foreword
Loop impedance see Earth fault loop impedance

M
Manual, operation and
  maintenance Foreword
Maximum demand Appx A
Mechanical maintenance,
  switching for 5.2
Meter 2.2.2
Meter tails 2.2.3
Mineral insulated cable Table C1
Minor Works Certificate Appx G

N
Nominal voltage 1.1ii; 6.4
Non-sheathed cables 3.4.1.1
Non-standard colours 6.13
Notices 6

O
Off-peak supplies 1.1a
Ohmmeter 10.3.1
Outbuildings 1.1a
Overhead lines/wiring 3.7.2.1; Appx D
Overload protection 3.2
Overvoltages 3.7.2

P
Part P 1.2.1
PELV 3.4.3; 7.3.1v; 7.3.2vi; 9.2.2; 10.3.3v; Table 10.3.3
Periodic inspection Appx G
Phase sequence check 10.3.8
Photovoltaic systems 6.16
Plastic services 4.5; 4.8
Polarity testing 9.3.1; 10.2; 10.3.4
Portable generators 2.4
Prospective fault current 7.2.7
  measurement 9.3.1; 10.2.2; 10.3.7
Protective device, choice 7.2.7
Proximity to communications cables 7.4.2

R
Radial circuits Table 7.1(i); Appx H
  testing 10.3.1
Rated short-circuit capacities Table 7.2.7(i)
RCBOs 3.6.3c; 7.2.6
RCDs 2.2.5; 3.6; 7.2.4; 9.2.2d
  diagram of operation Fig 11.0
  integral test device 11.6
  labelling 6.12
  multipole 11.7
  omission of 3.6.2; 7.2.5
  requirements 7.2.5
  testing 11
Reports Appx G
Residual current devices see RCDs
Resistance of conductors Appx I
Ring circuits Table 7.1(i); Appx H
  spurs 7.2.2; H2.4
  testing 10.3.2

S
Schedules 9.1; Appx G
Scope Preface; 1.1
Selection
  cables Appx C
  devices for isolation etc. Appx J
SELV 3.4.3; 7.3.1v; 7.3.2vi; 9.2.2; 10.3.3v; Table 10.3.3
Separation of gas pipework 2.3; Fig 2.3; 7.4.3
Short-circuit capacity Table 7.2.7(i)
Shower 7.2.5; 8
Socket-outlets
  final circuits with high PE current 7.5.3
  general 7.2.2; Appx H
  minimum number 1.2.2; Table H7

Index

Socket-outlets – *contd*
   protection by RCD   3.6.1ii
SPDs   3.7
   decision flow chart   Fig 3.7.2.2
   conductor critical length   3.7.5
   connection methods   3.7.6
   selection   3.7.4; Fig 3.7.4
   types   3.7.3
Split consumer unit   Figs 3.6.3(i)/(ii)/(iv)/(v)
Spurs   7.2.2; H2.4
Standard circuits   Appx H
Stud walls   Tables 7.1(ii), 7.2(iii), F6
Supplementary equipotential
   bonding   4.6–4.8; Table 4.6
Supplier (definition)   1.1
Supply
   frequency   1.1i
   nominal voltage   1.1ii
Support, methods of   Appx D
Surge protective devices   *see* SPDs
Surges
   lightning   3.7.1; 3.7.2
   switching   3.7.1
Switchgear   5.1.2; 6.3
Switching   5

T

Tails (consumer)   2.2.3; Figs 2.1(i)–(iii)
Telecommunications lines   3.7.1
Testing   9
   instruments   10.1
   procedures   10.3
   results schedule   Appx G
   sequence   10.2
   checklist   9.3.1
Thermal insulation   Table 7.1(iii); Appx F
Thermoplastics/thermosetting,
   applications   Table C1
TN system
   conduit installations   3.6.3a

disconnection times   3.5.2
TN-S system
   earthing arrangement   Fig 2.1(ii)
   typical external impedance   7.1
TN-C-S system
   earthing arrangement   Fig 2.1(i)
   typical external impedance   7.1
Toxic substances   1.2.1
Trunking
   capacities   Appx E
   supports   Table D4
TT system
   disconnection times   3.5.3
   conduit installations   3.6.3b
   earthing arrangement   Fig 2.1(iii)
   general   7.1; 7.2.6; 10.3.5.1
Two-way switching   7.4.4; Fig 7.4.4

U

Underfloor heating   8.3
Unexpected nominal voltage,
   warning of   6.4

V

Ventilation   1.2.1
Voltage drop
   calculation   Appx F
   general   7.1; 7.2.3; Table 7.1(i) note 1
   inspection   9.2.2
   verification   10.3.10

W

Walls and partitions   7.3.2
Water heaters   Appx H

Z

Zones
   bathrooms   8.1
   cables in walls   7.3.2v; 9.2.2

IET Wiring Regulations and associated publications

The IET prepares regulations for the safety of electrical installations, the IET Wiring Regulations (BS 7671 *Requirements for Electrical Installations*), which are the standard for the UK and many other countries. The IET also offers guidance around BS 7671 in the form of the Guidance Notes series and the Electrician's Guides as well as running a technical helpline. The Wiring Regulations and guidance are now also available as e-books through Wiring Regulations Digital (see overleaf).

IET Members receive discounts across IET publications and e-book packages.

Requirements for Electrical Installations, IET Wiring Regulations 17th Edition (BS 7671:2008 incorporating Amendment No. 1:2011)
Order book PWR1701B Paperback 2011
ISBN: 978-1-84919-269-9 **£80**

On-Site Guide BS 7671:2008 (2011)
Order book PWGO171B Paperback 2011
ISBN: 978-1-84919-287-3 **£24**

IET Guidance Notes

The IET also publishes a series of Guidance Notes enlarging upon and amplifying the particular requirements of a part of the IET Wiring Regulations.

Guidance Note 1: Selection & Erection, 6th Edition
Order book PWG1171B Paperback 2012
ISBN: 978-1-84919-271-2 **£32**

Guidance Note 2: Isolation & Switching, 6th Edition
Order book PWG2171B Paperback 2012
ISBN: 978-1-84919-273-6 **£27**

Guidance Note 3: Inspection & Testing, 6th Edition
Order book PWG3171B Paperback 2012
ISBN: 978-1-84919-275-0 **£27**

Guidance Note 4: Protection Against Fire, 6th Edition
Order book PWG4171B Paperback 2012
ISBN: 978-1-84919-277-4 **£27**

Guidance Note 5: Protection Against Electric Shock, 6th Edition
Order book PWG5171B Paperback 2012
ISBN: 978-1-84919-279-8 **£27**

Guidance Note 6: Protection Against Overcurrent, 6th Edition
Order book PWG6171B Paperback 2012
ISBN: 978-1-84919-281-1 **£27**

Guidance Note 7: Special Locations, 4th Edition
Order book PWG7171B Paperback 2012
ISBN: 978-1-84919-283-5 **£27**

Guidance Note 8: Earthing & Bonding, 2nd Edition
Order book PWG8171B Paperback 2012
ISBN: 978-1-84919-285-9 **£27**

Electrician's Guides

Electrician's Guide to the Building Regulations Part P, 2nd Edition
Order book PWGP170B Paperback 2008
ISBN: 978-0-86341-862-4 **£22**
To be updated 2013

continues overleaf ▶

Electrical Installation Design Guide, 2nd Edition
Order book PWGC171B Paperback 2013
ISBN: 978-1-84919-657-4 **£35**

Electrician's Guide to Emergency Lighting
Order book PWR05020 Paperback 2009
ISBN: 978-0-86341-551-7 **£22**

Electrician's Guide to Fire Detection and Alarm Systems
Order book PWR05130 Paperback 2010
ISBN: 978-1-84919-130-2 **£22**

Other guidance

Code of Practice for In-service Inspection and Testing of Electrical Equipment, 4th Edition
Order book PWR02340 Paperback 2012
ISBN: 978-1-84919-626-0 **£40**

Commentary on IEE Wiring Regulations (17th Edition, BS 7671:2008)
Order book PWR08640 Hardback 2009
ISBN: 978-0-86341-966-9 **£65**

Electrical Maintenance, 2nd Edition
Order book PWR05100 Paperback 2006
ISBN: 978-0-86341-563-0 **£40**

Electrical Craft Principles, Volume 1, 5th Edition
Order book PBNS0330 Paperback 2009
ISBN: 978-0-86341-932-4 **£25**

Electrical Craft Principles, Volume 2, 5th Edition
Order book PBNS0340 Paperback 2009
ISBN: 978-0-86341-933-1 **£25**

For more information and to buy the IET Wiring Regulations and associated guidance, visit www.theiet.org/electrical

Wiring Regulations Digital

No extras. No fluff. We've made the IET Wiring Regulations simple to download to your computer, tablet or smartphone. Start with one of our new package options or pick and mix the Guidance Notes and supporting e-books one at a time* as it suits you, using VitalSource Bookshelf®.

No more flipping through bulky hard copies looking for crucial information – a quick search brings up the facts you need, instantly. Our portable, electronic format means you can carry your e-books with you on the job, make notes and copy important text into other documents.

Purchase new titles as they are published giving you quick and easy access to the most up-to-date guidance.

Visit www.theiet.org/digital.regs for more information and your free trial.

* BS 7671 must be purchased with the On-Site Guide – buy together for £98.70 (Inc. VAT)

Order Form

How to order

PHONE:
+44 (0)1438 767328

FAX:
+44 (0)1438 767375

POST:
The Institution of
Engineering
and Technology,
PO Box 96,
Stevenage
SG1 2SD, UK

ONLINE:
www.theiet.org/electrical

INFORMATION SECURITY
Please do not submit your form by email. The IET takes the security of your personal details and credit/debit card information very seriously and will not process email transactions.

Postage within the UK. Outside UK/Europe) add £5.00 for first title and £2.00 for each additional book. Rest of World add £7.50 for the first and £2.00 for each additional title. Books will be sent via airmail. Other rates are available on request, please call +44 (0) 1438 767328 or email sales@theiet.org for rates.

Member Discounts: These cannot be used in conjunction with any other IET discount offers.

GUARANTEED RIGHT OF RETURN:
If you are unsatisfied, you may return books in new condition within 30 days for a full refund. Please include a copy of your invoice.

DATA PROTECTION:
The information that you provide to the IET will be used to ensure we provide you with products and services that best meet your needs. This may include the promotion of specific IET products and services by post and/or electronic means. By providing us with your email address and/or mobile telephone number you agree that we may contact you by electronic means. You can change this preference at any time visiting www.theiet.org/my.

Details

Name:
Job Title:
Company/Institution:
Address:
Postcode: Country:
Tel: Fax:
Email:
Membership No (if Institution member):

Ordering information

Quantity	Book No.	Title/Author	Price (£)
		Subtotal	
		Member discount **	
		+ Postage /Handling*	
		Total	

Payment methods

☐ By **cheque** made payable to The Institution of Engineering and Technology

☐ By **credit/debit card**: ☐☐☐☐ ☐☐☐☐ ☐☐☐☐ ☐☐☐☐

☐ Visa   ☐ Mastercard   ☐ American Express   Maestro Issue No: _____

Valid from: ☐☐ ☐☐   Expiry Date: ☐☐ ☐☐   Card Security Code: ☐☐☐☐
(3 or 4 digits on reverse of card)

Signature _____ Date _____

Cardholder Name:
Cardholder Address:
Town: Postcode:
Country:
Tel:
Email:

The Institution of Engineering and Technology is registered as a Charity in England & Wales (no 211014) and Scotland (no SC038698). Michael Faraday House, Six Hills Way, Stevenage, SG1 2AY

Dedicated website

Everything that you need from the IET is now in one place. Ensure that you are up-to-date with BS 7671 and find guidance by visiting our dedicated website for the electrical industry.

Catch up with the latest forum discussions, download the Wiring Regulations forms (as listed in BS 7671) and read articles from the IET's **free** Wiring Matters magazine.

The IET Wiring Regulations BS 7671:2008 (2011) and all associated guidance publications can be bought directly from the site. You can even pre-order titles that have not yet been published at discounted prices.
www.theiet.org/electrical

Membership

Passionate about your career? Become an IET Member and benefit from a range of benefits from the industry experts. As co-publishers of the IET Wiring Regulations, we can assist you in demonstrating your technical professional competence and support you with all your training and career development needs.

The Institution of Engineering and Technology is the professional home for life for engineers and technicians. With over 150,000 members in 127 countries, the IET is the largest professional body of engineers in Europe.

Joining the IET and having access to tailored products and services will become invaluable for your career and can be your first step towards professional qualifications.

You can take advantage of...

- a 35% discount on BS 7671:2008 (2011) IET Wiring Regulations, associated guidance publications and Wiring Regulations Digital
- career support services to assist throughout your professional development
- dedicated training courses, seminars and events covering a wide range of subjects and skills
- an array of specialist online communities
- professional development events covering a wide range of topics from essential business skills to time management, health and safety, life skills and many more
- access to over 100 Local Networks around the world
- live IET.tv event footage
- instant on-line access to over 70,000 books, 3,000 periodicals and full-text collections of electronic articles through the Virtual library, wherever you are in the world.

Join online today: **www.theiet.org/join** or contact our membership and customer service centre on +44 (0)1438 765678.